D1600017

THE CHIMPANZEES WHO WOULD BE ANTS

The Evolutionary Epic of Humanity

"Work fascinates me. I could lay here and watch ants for hours. Our intelligence, if combined with their organized, selfless industry, could take over the world."

THE CHIMPANZEES WHO WOULD BE ANTS

The Evolutionary Epic of Humanity

Russell Merle Genet

NOVA SCIENCE PUBLICATIONS
Commack, NY

Assistant Vice President/Art Director: Maria Ester Hawrys

Office Manager: Annette Hellinger

Graphics: Frank Grucci

Acquisitions Editor: Tatiana Shohov

Book Production: Ludmila Kwartiroff, Christine Mathosian,
Maria A. Olmsted and Tammy Sauter

Editorial Production: Susan Boriotti

Circulation: Cathy DeGregory and Maryanne Schmidt

Library of Congress Cataloging-in-Publication Data

Genet, Russell
The chimpanzees who would be ants / by Russell Genet
       p.   cm.
    Includes bibliographical references (p.) and index.
    ISBN 1-56072-522-2
    1. Evolution (Biology)--Philosophy. 2. Life (Biology)-Philosophy. 3. Social evolution—
Philosophy. I. Title.
QH331.G435   1997   97-42071
576.8'01—dc21   CIP

DEDICATION

This book is appreciatively dedicated to

David H. Poore

My seventh grade teacher who, having whetted my intellectual appetite, shrewdly hinted that humanity's accumulated knowledge was patiently waiting for me at the local, well-stocked public library.

Richard and Cheryl Fallick

My childhood confidants who demanded exhaustive reports as I read through philosophy, history, and science in my teenage quest for human meaning in a vast universe.

and

Elizabeth Gillis Genet and Harriet Whiting Gillis

My mother and grandmother, missionaries who kindly suggested that God would never condemn a young lad's honest search for truth (and never failed to take me to the library).

CONTENTS

PREFACE
Why Science is Perverse

As Hitler's blitzkrieg smashed Poland, an Austrian, Nobel prize-winning physicist Edwin Schrodinger, made good his escape to neutral Ireland and academic freedom at Trinity College, Dublin. While the world slugged it out, Schrodinger, the inventor of quantum-wave mechanics, applied his physical insights to an alien field, biology. In 1944 he gave a public lecture at Trinity, *What Is Life?* His keen physical insights into how life *had* to be different from non-life inspired Francis Crick to find the physical basis for life -- DNA. In a Preface to the published version of his lecture, Schrodinger captured both my aim in writing this present book and the difficulty I have had in doing so.

"A scientist is supposed to have a complete and thorough knowledge, at first hand, of *some* subjects and, therefore, is usually expected not to write on any topic of which he is not a master. This is regarded as a matter of *nobless oblige*. For the present purpose I beg to renounce the *noblesse*, if any, and to be freed of the ensuing obligation. My excuse is as follows:

We have inherited from our forefathers a keen longing for unified, all-embracing knowledge. The very name given to the highest institutions of learning reminds us that from antiquity and throughout many centuries, the *universal* aspect has been the only one to be given full credit. But the spread, both in width and depth, of the multifarious branches of knowledge during the last hundred odd years has confronted us with a queer dilemma. We feel clearly that we are only now beginning to acquire reliable material for welding together the sum-total of what is known into a whole; but, on the other hand, it has become next to impossible for a single mind to fully command more than a small specialized portion of it.

I can see no other escape from this dilemma (lest our true aim be lost forever) than that some of us should venture to embark on a synthesis of facts and theories, albeit with second-hand and incomplete knowledge of some of them, and at the risk of making fools of themselves."

My aim is more ambitious than Schrodinger's. Rather than applying a few ideas from one field of science to a single question in another, I plan to utilize numerous ideas from the natural and social science to answer three of humanity's deepest questions: Who are we? How did we come to be? and What is our fate? Unlike Schrodinger, I offer nothing original. My own ideas are purposely excluded. My intention is to be the integrating (albeit uninvited) spokesperson for a wide range of scientists, each expert in their own field.

Science is perverse. On minutia with only the narrowest of interest, science waxes eloquent. On broad human questions everyone cares about, science is usually silent. When it does offer opinions they are often in disagreement with each other. Why is this?

First, it's not Johnny-come-lately science's job to answer humanity's eternal questions. This is the function of those noble enterprises of the human spirit; religion, philosophy, literature, and the arts. They, not science, have evolved over thousands of years to provide meaning to daily lives, to smoothly harness our wild animal spirits to the regimenting demands of civilization, to provide intuitively pleasing and humanly gratifying answers to our deepest, most persistent questions. Science is wisely reluctant to challenge these other ways of knowing on their home turf.

Second, in protecting its paragon of virtue, objectivity, science eschews appeals to human desire, needs, and preferences. Banned from the Halls of Science are private feelings, cultural works, and all that which makes us human (except our curiosity and occasional desire for logical consistency). Humans, until proven otherwise, are no different than ten million or so other animal species on Earth. Science does its level best to be non-human. It should come as no surprise that science has the greatest difficulty in answering deeply human questions.

Third, science specializes in reducing knowledge into ever smaller disciplines until, at the frontiers of science, there are thousands of minute but workable specialties. Such reductionism has paid handsome dividends -- a fact not lost on scientists themselves. In actuality there is no *science*, per se, but a number of totally independent, self perpetuating and, at best loosely cooperating *sciences*. You won't find a single Department of Science at major universities. Instead there are numerous independent departments from Anthropology to Zoology. Each has its own language, viewpoint, textbooks, journals, and meetings. Zoologist speak to other zoologists world wide, but rarely talk to sociologists at the same university.

Finally, when science does offer ideas on major human questions, they usually contradict each other because the viewpoints of the independent sciences are so divergent. From their God-like perspective, physical scientists consider humanity just one of ten million animals inhabiting the surface scum of a third-rate planet in a remote corner of an ordinary galaxy. Biologists are less fortunate. Forced to consider life on a single planet, they shrink from grandiose, universe-wide conclusions. Still, with millions of species to consider over billions of years, biology is rich with cross-species comparisons and evolutionary trends. Biologists don't have to ruin their objectivity by considering that one troublesome species, *Homo sapiens*. Social scientists must not only be content with this single species, but the past few thousand years. They rarely draw cross-species or evolutionary conclusions. Yet they don't complain as they consider both human culture and the last few millennia as phenomena unto themselves. Unrestrained by grubby physical or biological constraints, human culture and civilizations have blossomed to become the most interesting phenomena of all. With such widely differing views, it is little wonder that the natural and social scientists disagree when it comes to humanity.

This being the case, why even attempt answers that cross the natural and social sciences in an interdisciplinary manner? Why not forget about answering broad human questions from a unified scientific viewpoint? Are there any compelling reasons for doing so beyond satisfying Schrodinger's "keen longing for unified, all-embracing knowledge," beyond scratching an intellectual, academic itch? Two reasons suggest themselves: First, biology, especially ecology, is no longer understandable without considering that recent major biospheric factor, civilized humanity. Second and conversely, social scientists can no longer assume that humanity is independent of biology. Our human actions and social institutions are and must be increasingly concerned with biological, with planetary constraints. To understand our current human predicament, social scientists need to appreciate how, evolutionarily, it came about.

It is our good fortune that "reliable material for welding together the sum-total of what is known into a whole" is, a half-century after Schrodinger's talk, more plentiful then ever. Scientific knowledge has mushroomed. Like a child with a new set of blocks, how can we resist trying to put the pieces together? I invite you to join me as, taking Schrodinger's advise, we attempt a synthesis, as we throw reductionist science in reverse so that, together, we will come to know less and less about more and more. We risk making fools of ourselves, but we have a planet to save.

ACKNOWLEDGMENTS

True Confessions

I confess. I borrowed it all, science's unified view of humanity. In this age of specialization it is impossible to become expert across science, yet only a fool would write on something they were not expert. Not being expert, I borrowed it from experts. Borrowing ideas isn't easy. I didn't have the time to take dozens of university courses in biology, the social sciences, and history. Nor was I patient enough to wade through dozens of strangely worded, expensive textbooks, each loaded down with tons of unwanted trivia. So I borrowed it from low cost, easy-to-read science "trade" books. Written by eminent scientists for the edification of the unwashed masses, they were easy pickings. The only real work was finding the most authoritative, eloquent, and easily understood books. They are all dog-eared now, full of notes scribbled in the margins. I am deeply indebted to these literate scientists.

With key ideas in hand, the first draft followed in due course. Not being expert myself, how would I know if I got it right? I was stumped until it dawned on me that I could entice the experts themselves to correct my multitudinous mistakes. On the completion of each draft, of which there were four, individual chapters (and occasionally entire manuscripts) were sent to dozens of experts for criticism. They were always willing to help, perhaps unable to resist participating in my bold attempt to put the pieces together (not to mention protecting the good name of their individual specialties from a foolish generalist). In a cruel stroke of revengeful genius, I even sent draft chapters to some of the very scientists whose delightful books were my source material. My expert advisers included John Alcock, Connie Barlow, Thomas Berry, Henry Byerly, Roger Caldwell, Tyrone Cashman, Patric Colburt, Dwight Collins,

Michael Corballis, George Cowgill, Russell Davis, Richard Eaton, Lawrence Edwards, Marcus Ford, Donald Hayes, Richard Healey, Kenneth Teisell, James Liu, Peter Manly, Mary Marzke, William McGrew, Mary Ellen Morbeck, Leanne Nash, Edward Olson, Ronald Olowin, Charles Redman, Steven Rissing, Loyal Rue, Jean Schmidt, Richard Shand, Steven Stanley, Thomas Suddendorf, Lloyd Swenson, Brian Swimme, Mary Evelyn Tucker, Luiz Clementino Vivacua, Bruce Weber, Andy Wilson, James Whetterer, and Diana Wheeler. They not only pointed out numerous errors, but assigned further books for me to read. By the fourth draft the number of critical comments and their severity had markedly decreased. Their continued presence, however, suggests either that scientists never reach agreement with each other or, more likely, that I still have much to learn. My advisers are not responsible for the errors and omission which remain, nor the way I merged the various disciplines.

After much expert help and two drafts, I bounced my manuscript off a few personal friends and local high school students -- stand-ins for my intended audience. Did they understand it? Was it fun to read? My "beta test" readers included Louis Boyd, Cheryl Davidson, David Easingwood, Celeste Espinoza, Richard Fallick, Ismari Garcia, Hal Geltmacher, Gloria Genet, my children (David, Michael, Karen, Shirly, and Russell Genet), Adriana Gerardo, Julia Hayes, Art Hoag, Danny Mandel, John Molinder, Edward Olson, Michelle Rabago, Mark Trueblood, Scott Vandervet, and Bernard Wells. I am indebted to them for their suggestions. They are not responsible for remaining incomprehensibility. I am indebted to Pat Trueblood and Elaine Evans who each proofed a draft, and to Connie Barlow who injected much-needed liveliness.

Anastasia Moreno, a student at the University of Arizona, kindly drew up a dozen cartoons based on various mental pictured I had formed. We chose the best four of page for this book's frontpiece and section dividers.

Finally, I want to thank my wife, Joyce, for patiently enduring three "winters" on the beach at Awaroa, New Zealand (actually it was summer), and three summers in the Arizona mountains. Accompanying a writer is a tough job, but someone has to do it!

I. WHO ARE WE?

Revealing perspectives from other life

"Humans are so conceited. In this book they claim that they are the only species to make tools, the only species with language, and get this, the only self conscious species. Who do they think they are?"

CHAPTER 1

CHIMPANZEES
Masters of Tools and Machiavellian Intrigue

Who are we? Are we just another animal? Or are we, as many believe, so unique that we belong in a separate category of life with ourselves as its sole occupant? The first section of this book will address these questions.

In so doing we will adhere to the spirit of science, to its objectivity and disinterest. We'll avoid allowing our own humanity to influence our answers. In the past, we humans (yes, even scientists) falsely believed that the Earth was the center of the universe. Then it was the sun. And then our own Galaxy. Obviously we will have to constantly guard against giving ourselves a falsely central or privileged position, in crediting ourselves with a uniqueness that we don't deserve.

Perhaps the best way to avoid human bias in answering these questions is to assume, at the outset, that we humans are just another of the approximately ten million species of animals currently inhabiting the surface of Earth; that we are not particularly unique when compared with the normal differences between animals. "No out of the ordinary uniqueness" will be our default and final answer unless it can be objectively and unmistakenly shown that we are inarguable unique in some clearly non-anthropocentric, major way. The burden of proof will fall to those who claim human uniqueness. The benefit of any doubt will be given to those suggesting we are just another animal. As objective "scientists," our goal will be to "explain away" apparent uniqueness. If some human feature appears unique, we will consider it "shot down" if we can find any life that does essentially the same thing.

After going through the process of shooting down supposedly unique human features by way of examples from other life, our answer should, logically, take one of two forms: (1) No major features will be left unexplained, thus making us just another animal; or (2) A significant residue of human uniqueness (well beyond the normal differences between species) will remain, and we will be forced, however reluctantly, to concede that we aren't just another animal.

Biologists have found that closely related species are usually quite similar. This similarity is due, logically enough, to descent from a recent common ancestor. By considering our nearest living relatives -- especially their behavior in the wild -- we hope to effectively (i.e. scientifically) demolish most claims of human uniqueness. The animals that are our closest living relatives are two species of chimpanzees. Perhaps, as suggested by the catchy title of Jared Diamond's book, we are just *The Third Chimpanzee*.

The species *Pan troglodytes*, the "common" chimpanzee, comes to mind when chimpanzees are mentioned, but there is another species, *Pan paniscus*, known as the bonobo. Sometimes called the pygmy chimpanzee, the bonobo is only slightly smaller than its better-known relative. What is striking is the bonobo's human-like appearance. Unlike *troglodytes'* projecting snout, *paniscus* is flat faced, maintains a more upright stance, and engages in face-to-face sex.

Chimpanzees primarily eat fruit. It makes up some two-thirds of their diet. They also eat young leaves, seeds, and insects. The common chimpanzees eat the meat of small animals which they cooperatively hunt. About half their waking hours are spent feeding, the rest traveling to new trees, grooming, and other activities. Muscular (dangerously stronger than humans), chimpanzees climb trees with ease and are comfortable traveling a hundred feet off the ground, although they prefer to travel long distances by knuckle walking on the ground. Bonobos are less violent than the common chimpanzees and do not hunt. Anatomically, bonobos appear to be more like "Lucy," the fossil hominid found by Donald Johanson, than are the common chimpanzees.

During the Ice Ages, vast amounts of water were locked up in huge ice sheets. The glaciers reflected much of the sunlight back out into space. As the Earth's climate turned dryer and cooler, the African rain forests shrank to a fifth of their current area. Only a few isolated pockets remained. The chimpanzees conservatively stayed with the jungles, although it was a close call for them as the jungles almost disappeared. Today, chimpanzees are being threatened again, and are only found in a few remaining sections of relatively undisturbed forest in Central and West Africa. Chimpanzees never left the Garden of Eden.

That the two species of chimpanzee are our closest relatives is no longer in doubt. DNA analysis shows that we three descended from a common ancestor who lived some six million years ago. We share 98% of our genes with chimpanzees. Gorillas and orangutans, our next closest relatives, are slightly more DNA distant. Put another way, humans, not gorillas or orangutans, are the closest relatives of chimpanzees and bonobos. When Linnaeus, the Swedish classifier of life, lumped great apes (chimpanzees, gorillas, and orangutans) together but placed humans in a separate category, he made a zoological classification error. We are a great ape and, perhaps, should be a third species of chimpanzee, *Pan sapiens*.

Anatomically and genetically, there is no longer any question that we are, in essence, a third species of chimpanzee. Most scientists, however, still assume that behaviorally we are worlds apart from our two sister species. Decades ago, Louis Leakey realized that our understanding of chimpanzee behavior, based on observations made in zoos and other captive conditions, was inadequate as a basis of comparison to our own behavior. True understanding would only come, he reasoned, by observing how our closest relatives actually lived in the wild. To this end he arranged, in the early 1960s, for Jane Goodall to observe a group of wild chimpanzees in East Africa. Observe she did, for over three decades.

Goodall was not trained as a scientist before beginning her observations. She was not aware that it was scientifically improper to give individual animals names, let alone invoking human-style emotions when explaining their behavior. For most animals, such anthropomorphism (impugning of human motivations) would indeed have been a serious scientific error. In the case of chimpanzees, however, it was a fortuitous stroke of good luck. For animals other than chimpanzees, our natural tendency is to credit them with too many human attributes. With chimpanzees our tendency had been to distance ourselves from them more than we should. As subsequent research has demonstrated, with chimpanzees one is safest in assuming they are just like us.

One of humanity's longest-standing claims to uniqueness is that only we are self-conscious. Jane Goodall's field observations suggested that chimpanzees are self-conscious too. For those desiring a more formal scientific study under controlled conditions, Gordon Gallup conducted in the late 1960s his famous mirror-and-red-spot experiment. Various animals were provided mirrors. Those that could view their own image without panic were enlisted in Gallup's clever experiment. While anesthetized, a harmless red spot was surreptitiously painted on their foreheads. On awakening and looking in a mirror, individuals of only three species immediately reached to their foreheads to feel the strange red spot they observed on themselves in the mirror. These species were chimpanzees, orangutans, and humans. (A sign-language trained captive gorilla, Koko, later did the same thing.) Gallup suggested that it was not just a case of apes figuring out how a mirror works. Several species of monkeys are able to do this, using mirrors to find hidden food. Rather, apes identify themselves in mirrors because, like us, they have a concept of self.

Jane Goodall's classic book, *In the Shadow of Man*, based on her first decade of common chimpanzee observations, reads like a biography of the Roosevelts or Kennedys -- one soon forgets that the characters are not human. The strong family bonds, lifelong friendships, rivalries, and political intrigues are all endearingly human. Sadly, Goodall's early portrait of happy, playful, generally loving chimpanzees does not prevail to this day. Chimpanzees are not entirely good-natured. Jane Goodall's continued observations eventually revealed a darker side: the psychotically jealous baby killer; the cold, ruthless killing, one by one, of a breakaway subgroup. Her second popular book, *Through a Window: My Thirty Years With the Chimpanzees of Gombe*, reveals these unpleasant but hauntingly human facets of chimpanzee behavior. The common chimpanzees now appear to be more like a Mafia family with occasional murders and sadistic beatings contrasting with their normally warm, affectionate nature.

Bonobos have now also been observed extensively, both in their natural setting south of the Congo river in Central Africa and in a large colony at the San Diego Zoo. Bonobos look strikingly human because, like ourselves, they have retained many juvenile features into adulthood; the juvenile features of our common ancestor. Unlike common chimpanzees, bonobo sexes are of nearly equal size, indicative of their egalitarian society. Following the dictate "make love, not war," bonobo political life has a decided, and to us prudish humans, embarrassingly blatant sexual slant. Sex is the grease that smooths the workings of their peaceful societies.

Frans de Waal observed, recorded, and analyzed the social interactions in a sizable colony of common chimpanzees that lived relatively undisturbed on their own island in a zoo in Belgium. He found that the social interactions between chimpanzees were politically astute. Brute strength took a back seat to political finesse. These various observations of chimpanzees suggested that their large brains resulted from a within-species arms race of sorts, a game of intricate social chess. Chimpanzees had evolved Machiavellian intelligence. Although arranged in a hierarchical order, the rank of individual chimpanzees does not correlate directly with physical power. The alpha male is not always the largest or strongest. A network of relatives and friends are built up, often over a lifetime. Who you know counts in chimpanzee society.

The alpha male is an astute politician, currying favor with female supporters by playing with their babies and even kissing them (Presidential politics has been around for a long time). Chimpanzees are good at remembering who owes them favors, and to whom they owe favors. Keeping track of constantly shifting alliances in a group of forty or fifty chimpanzees is a major intellectual accomplishment, but one they have evolved large brains to handle. Like us, they love soap operas.

Chimpanzees are, as Franz de Wall suggests in the title of the his latest book, inherently *Good Natured.* They are relatively nonviolent, faithful, caring, and loving to family and friends. They usually settle conflicts within their community by prompt reconciliations accompanied by grooming, hugs, make-up kisses and, in the case of the bonobos, sex.

Chimpanzees have well-developed mental models of long-term personalities and short-term emotions and motives of other chimpanzees in their group, keeping careful account of the complex and ever-shifting interrelationships among those they live with day-to-day. Chimpanzees are not only self-conscious, they are always projecting ahead. "If I do this, he will do that, and then she will join..." These are mental models of social cause and effect.

Goodall observed, early on, a behavior heretofore thought to be uniquely human: the manufacture and use of tools. Tool-making chimpanzees at Gombe break off long stems of grass, clip off side leaves and excess lengths, and then insert these tools into the entrances of termite nests. Pausing to allow the termites to attack this intruder into their nest, they slowly and carefully withdraw their tools, now teeming with protein. This tool-making, tool-using tradition is passed from one generation to the next. Subsequent observations by other primatologists have revealed that tool traditions are not unique to Gombe, although specifics vary from tribe to tribe.

Chimpanzees in different groups use leaves as tools in various ways. Leaves are used as umbrellas, sponges for water, or as personal hygiene napkins. The most amazing cultural tradition of all is that of the nut-cracking chimpanzees of the Tai forest in West Africa. Observed by Christophe Boesch, a Swiss primatologist, these common chimpanzees eat a plentiful nut whose shell is so hard that it can only be broken open by using a stone hammer. The stone must precisely strike a nut carefully held with fingers in the slight depression of a wooden anvil log. Avoiding smashing the nut to an inedible pulp of shell and meat, not to mention smashing fingers, takes great skill. It can be and is learned, but it takes several years and hundreds of practice sessions before young chimpanzees finally get it right.

Mothers patiently teach their children to crack nuts. First they leave nuts and a hammer stone near an anvil to play with. As their children get older, they encourage them to try cracking nuts, supplying them with nuts. If an inappropriate technique is used, mother will interrupt, slowly and carefully demonstrating the correct technique, and then letting her child carry on. Learning to crack these nuts is worthwhile. In season, they provide over half the chimpanzee's calories.

So, how long have the Tai chimpanzees been practicing nut cracking? Is it truly a cultural tradition? Boesch points out that in the Ivory Coast, chimpanzees on one side of the Sassandra river crack nuts, while those on the other side, where the nuts are just as numerous, don't. During the last Ice Age, about 17 thousand years ago, the jungle was greatly reduced in area, breaking up into isolated pockets. Apparently the nut-cracking tradition spread from a single pocket of the jungle as it expanded again, thanks to improved weather. This spread was stopped, however, by the Sassandra river. How much further back in time this nut-cracking tradition goes is not known. Perhaps for hundreds of thousands of years.

Chimpanzees have now been studied extensively under artificial but pleasant and enriching environments. Washoe was raised by the Gardeners in their home in a manner similar to a human child. Because the vocal tracts of chimpanzees are not capable of human speech, the Gardeners taught Washoe American Sign Language. Washoe was able to learn about a hundred signs; mainly objects and verbs. To the Gardeners it was clear that Washoe had acquired the capability of speech, but a few vocal critics (no pun intended) remained dubious.

At the Yerkes Primate Center, near Atlanta, primatologists Sue Savage-Rumbaugh and Diane Rumbaugh initiated language instruction of both common chimpanzees and bonobos. Instead of American Sign Language, they communicated by way of special computer keyboards and video displays, and also through spoken English (the chimpanzees responding via keyboard or action). Their efforts were, at first, only marginally successful. Then one day, while Sue Savage-Rumbaugh was patiently giving language lessons to a bonobo mother, the two-year-old son shocked Sue with a demonstration of his heretofore unsuspected skill of understanding spoken English. Kanzi tapped out meaningful answers on his mother's portable keyboard. Apparently Kanzi, like young human children everywhere, had easily learned a new language, while his mother, the intended student, had great difficulty in doing so. Kanzi went on to become an effective English communicator (as far as such things go among apes), learning 150 words by the age of six (comparable to the linguistic skills of a two-year-old human).

A bonobo for all seasons, Kanzi became the only ape known to fashion stone tools (as opposed to just selecting appropriate stones). Nicholas Tough, an anthropologist who specializes in making stone tools in a manner thought similar to that of our early ancestors, taught Kanzi the rudiments of this ancient art. Kanzi, with instruction and encouragement, is closer to being human than any animal has ever been.

Louis Leakey not only sent Jane Goodall to observe chimpanzees in the wild, but made it possible for Diane Fossey to observe gorillas and Birute Galdikas to observe orangutans. Leakey felt women made better, more patient observers than men. He thought they would appear less threatening to the male-dominated great ape societies. Diane Fossey's book, *Gorillas In the Mist*, was made into a movie not long after her untimely death. She was killed, perhaps in retaliation for her zealous protection of her gorillas against poachers.

Birute Galdikas, who observed orangutans deep in the Borneo Jungle for decades, recently wrote *Reflections of Eden*. Orangutans make some tools in the wild (and are very handy with some human tools), and share many other traits with other great apes and ourselves. It seems likely that much of what we only recently considered uniquely human is, in fact, traceable to the common ancestor of all great apes some twelve million years ago. Deep roots indeed. Louis Leakey's "three angels" (Goodall, Fossey, and Galdikas), along with dozens of hard-working primatologists from around the world, have revolutionized our view of humanity, giving it a distinctly more scientific, evolutionary flavor than before.

Are we a third species of chimpanzee? Anatomically and genetically the answer is an unreserved, Yes! Behaviorally, too, we are more similar to our chimpanzee kinsfolk -- and they to us -- than previously believed. We -- all three species -- are self-conscious, politically astute tool makers; social beings capable of symbolic communication.

FURTHER READING

Byrne, Richard, 1995, *The Thinking Ape: Evolutionary Origins of Intelligence.*
Fossey, Dian, 1983, *Gorillas In the Mist.*
Galdikas, Birute M. F., 1995, *Reflections of Eden: My Years With the Orangutans of Borneo.*
Goodall, Jane, 1977, *In the Shadow of Man.*
_____, 1990, *Through a Window, My Thirty Years with the Chimpanzees of Gombe.*
Griffin, Donald, 1992, *Animal Minds.*

Kaplan, Gisela, and Lesley Rogers, 1996, *Orangutans in Borneo*.

McGrew, William C., 1992, *Chimpanzee Material Culture: Implications for Human Evolution*.

McGrew, William C., Linda F. Marchant, and Toshisada Nishida, eds., 1996, *Great Ape Societies*.

Parker, Sue T., and Kathleen R. Gibson, 1990, *"Language" and Intelligence In Monkeys and Apes*.

Parker, Sue T., R. W. Mitchell, and M. L. Boccia, 1994, *Self-Awareness In Animals and Humans*.

Strum, Shirley C., 1987, *Almost Human: A Journey Into the World of Baboons*.

de Waal, Frans, 1982, *Chimpanzee Politics: Power and Sex Among Apes*.

_____, 1989, *Peacemaking Among Primates*.

_____, 1996, *Good Natured: The Origins of Right and Wrong In Humans and Other Animals*.

Wrangham, Richard W., W. C. McGrew, Frans de Waal, and Paul G. Heltne, eds., 1994, *Chimpanzee Cultures*.

CHAPTER 2

ANTS
The Perfect Little Communists

As loving, politically astute, clever, and self-conscious as chimpanzees and bonobos are, they don't engage in many of our civilized practices. Chimpanzees do not herd other animals, cultivate plants, or live in cities by the hundreds of thousands, let alone millions. Nor do they conduct highly organized warfare on a massive scale. Since we are obviously so different from our nearest relatives, can we, on this basis, safely conclude that we are not just another animal?

Not at all. Before we can draw such a conclusion, we need to consider other, more distantly related species. Perhaps some of them behave in these "human" ways? In fact, to find species that engage in the typically "civilized" behaviors mentioned above, we don't have to look very far. Our civilized analogs are not just in Africa. They are everywhere. They are ants.

Typical ants weigh about one ten-millionth as much as a human, but ants are about ten million times more numerous than we are. Thus their combined weight is roughly the same as our own. Planet wide, ants account for about 10% of the total land animal biomass, humans another 10%. The other ten million or so species of animals share the remainder.

Unlike chimpanzees, with whom we share a common ancestor about 6 million years ago, we would have to go back 600 million years, clear back to the earliest animals, to find the common ancestor of humans and ants. In the long course of evolution we should not be surprised if some of the problems involved in organizing thousands or even millions of animals to work together as a team has had similar solutions. Convergent evolution is a common theme throughout nature.

Ants are descendants of solitary wasps that lived over 100 million years ago. Solitary at first, ants led increasingly social lives, with adult ants eventually lavishing meticulous care on their young. By 30 million years ago, many ants had reached their present form and high degree of success. Ants, and the similarly social termites, account for almost three-quarters of the total insect biomass. In the Amazon jungle, they are an amazing one-third of the total animal biomass. From a biological perspective, the animals that weigh the most (have the most biomass), are the winners. What accounts for the spectacular biological success of the ants (and termites)? How were they able to banish less organized, less social insects to the periphery? Four mutually supportive reasons have been advanced.

Effective inter-ant communications is the first reason. Ants speak to each other chemically with a vocabulary of some ten to twenty words, depending on the species. Each chemical word is secreted by a separate gland controlled by the ant's tiny brain. There are ant messages for feed me, groom me, follow me, help me, alarm, and emergency evacuation, as well as for the identification of casts, larvae, and nest mates. These chemical messages are supplemented with sound and vibration. The ant's chemical words can, to some extent, be combined in different ways and in differing amounts to form various phrases. Meanings are explicit. Responses are entirely instinctual and uniform.

A few freeloading insects have "learned" (i.e. genetically evolved) to speak "Antease." Some beetles, for instance, have learned Antease for feed me. The ants, ignoring the huge, unantlike monstrosity saying the word, dutifully regurgitate the requested food. Different ant species secrete different pheromones (chemicals) from their various glands. Thus there are many different ant "languages," perhaps as many as there are different ant species. Freeloading beetles must specialize.

Chemical communication between members of the same species seems strange to us. With the exception of the persuasive messages of perfume, cologne, and our own natural pheromones that we have only lately been keen to investigate, human noses are relatively blind and deaf to chemical messages. Chemistry, however, is the most exploited mode of communication for life as a whole. Most organisms are simply too small to have eyes, ears, or other large communication devices, let alone the sizable brains required to operate them. Although ants speak in life's traditional tongue, they have raised chemical communication to new heights.

From our human perspective it is difficult to understand how, with only a score of words, ants can achieve the marvelous coordination they obviously do. Ant behavior is foreign to us because they have no high-level control. Queen ants are egg-laying machines, not coordinating bosses. Ants are completely decentralized. They respond to a combination of signals from other ants, the physical situation at hand, and their stored, mainly hard-wired behavioral programs. Such genetically predetermined behavioral repertoires appear odd to us. Our brains are extensively programmed by way of the bumps and bruises of our own lived experience.

Individual ants are less than brilliant learners. Ants as species, however, are excellent genetic learners. Over millions of generations, various ant species have "learned," i.e. have been genetically programmed, to accomplish, in decentralized ways with paltry vocabularies, amazing feats of large-scale coordination. The clever coordination of ants resembles intelligence. It is not. Individual ants are cogs in the ant-colony machine, genetically programmed to work harmoniously overall. It has been suggested, with some merit, that ants resemble silicon chips with six legs.

The second reason for the ant's success is their caste system of distinctly different but cooperating specialists. The largest ants are soldiers, followed by various types of workers. Larger workers forage for food and transport it to the nest; smaller workers tend the young. Soldiers and various workers are all sisters, as a single queen usually lays eggs for the entire colony.

A limited number of males are produced seasonally. Their sole task is to spread the genes of the colony to newly founded colonies. Few succeed. All quickly die.

Solitary insects, in contrast to the social, specialized ants, must accomplish all the different soldier, worker, and reproductive tasks with a single type of body. Nature's solution to such a wide variety of different demands must be a compromise. It is not surprising that solitary wasps usually lose if they try to compete directly with their evolutionary descendants, the highly efficient social ants.

Whether an ant egg becomes soldier, worker, or queen is not determined genetically. Ants are what they eat. Body type and career is determined by what food they are fed or what chemical conditioners spike their meals. This confers the advantage of individual specialization while allowing any given ant type to continue as a single species. Furthermore, it allows on-the-spot adjustments to be made in the relative proportions of soldiers and various workers as the colony matures or circumstances change. If, for instance, most of the soldiers are lost in a major war but few workers are, then new soldiers can be preferentially grown.

A third reason for the success of social ants *vis-a-vis* their solitary relatives is redundancy, a natural result of specialization and large numbers. A solitary insect must successfully accomplish all life's requisite tasks in series. One error, one lapse in any task, and the game of life is over. On the other hand, if one ant fails to do the proper thing, another will come along and set it right. Even if an ant dies trying, another will take its place and complete the task. The colony survives. Worker ants, including soldiers, are entirely expendable, and may only live for a few months before being replaced. Queen ants, and their colonies, may continue on for years, even decades. Such redundancy gives ants an edge over solitary rivals when they meet head on. Ants can afford to be ludicrously brave in combat. As Bert Holldobler and Edward O. Wilson colorfully suggest in their book, *Journey to the Ants*, ants "act like six-legged kamikazes."

The fourth and final key to the ant's success is, for many species, an elaborate and permanently constructed home, their architecturally clever "cities." As many subsequent generations of ants live in the same structures, they are able to maintain their claim to what is usually prime real estate. Obviously a case of animal property inheritance. Furthermore, ants are, over the generations, able to expand and refine their ancestral homes. The large size and layout of many ant colonies allows the occupants to regulate temperature and humidity. By elevating their main entrance, not only is rain run-off diverted, but a chimney is created that improves air circulation. Cool air drawn in via small holes around the nest's perimeter is expelled out the central, built-up chimney.

With so much going for social ants, we can't help but wonder why any solitary competitors remain. As anyone from a modern civilization would appreciate, large, complex societies do have their downside. Ant colonies are no exception. Solitary insects take advantage of the downside of such organized complexity by making do with limited, transient resources; living off the small crumbs left by the ants and other social insects at the periphery. Massiveness requires a large base of operations. Furthermore, ant colonies take time to develop their full potential. Colony relocation is difficult or impossible for many ant species. Once in place, however, ant colonies are, as Holldobler and Wilson put it, "ecological juggernauts."

Solitary insects, by definition, do not engage in organized warfare. Ants wouldn't need to either if their only competitors were solitary insects. Organized warfare is waged against other ants (and termites). In the words of Holldobler and Wilson, ants practice "relentless aggression, territorial conquest, and genocidal annihilation of neighboring colonies whenever possible. If ants had nuclear weapons they would probably end the world in a week."

Ant wars began when ants first became social, organized insects, and have been waged with increasing efficiency and intensity for the past 100 million years. As ants proliferated about the planet, so did the frequency and intensity of their warfare, fighting for increasingly scarce territory and food. The result was an arms race with ever more exotic offenses and countering defenses. Recent civilized humans aside, no animals come anywhere close to the large scale, highly evolved, and efficiently organized combat of the ants. As Eric Hoyt put it in *The Earth Dwellers*, "Other insects, birds, mammals, humans, all creatures of more recent origin, largely can only stand by and watch from the sidelines. A few conduct their own minor battles, of course, all of which are comparatively insignificant affairs with the exception of humans."

Ant offensive warfare is primarily chemical warfare; no milk-toast Geneva treaties for them! Nozzles spray poisonous chemicals on the enemy. Some ants have even evolved a specialized soldier caste into chemical bombs. The normally small poison gland has, in these soldiers, swelled to tank size. They lumber into the opposing ranks and then, quite literally, self explode, showering the enemy with poison. Eric Hoyt calls them "car bombers."

Ants are also masters of defense. In some species, large soldier ants serve as a living door to the nest, using their bodies to block access to all who lack the chemical password. Other species close the nest entrance with dirt at sunset or when threatened; they close the city gate.

Among ants, might makes right. Ant colonies that field the largest armies usually win, forcing smaller colonies to less desirable locations or annihilating them altogether. This is not always the case, however. Some species of ants with inherently small colonies have, over the generations, genetically evolved a clever strategy that allows them to hold onto their prime real estate. These species of ants maintain a rapid reaction force that, normally, does nothing but sit around and wait for action. If a single advanced scout from a nearby large-colony species of ants is detected, the rapid reaction force is summoned and the hapless scout is killed. But this is only the beginning. The entire area is then diligently searched for any other advance scouts and, if found, they too are summarily eliminated. As the large ant colonies never hear back from scouts sent into the small colony's territory, they have no reason to go there in force, and the smaller colony remains undetected. For these ants as with many human societies, constant vigilance is the price of freedom. The brilliant strategies and tactics of ant warfare are legion. We humans have little to add to this wisdom of the truly ancient.

While ant warfare is often directed towards other ant species (and the highly organized termites), pitched battles and even long, drawn out campaigns are also waged between colonies of the same species; warfare in the strictest sense of the word. The fierce Aztec ants of the Amazon are a case in point. Aztec colonies live in cecropias. A cecropia plant is the size of a large cactus. An Aztec colony protects their cecropia from other animals. In turn, the cecropia provides little food buds for the ants to eat. All is relatively peaceful until another Aztec colony moves into an already occupied cecropia. This triggers a no-holds-barred, take-no-prisoners, all-out campaign that may last for years until, at last, one colony is triumphant and the other totally vanquished. As Eric Hoyt suggests, it is a "multi year saga filled with forced evictions, numerous treacheries, blatant violence, bullying, occasional kidnaping, and pitched battles."

The Arizona desert, besides being occasional home to the author, is also home to honeypot ants. A typical colony consists of a queen, some 20 thousand sister workers and soldiers, and a couple of thousand honeypots. The honeypots spend their entire adult lives hanging from the nest's ceiling. During times of plenty they are fed large quantities of honey-like food, and become huge, distended, living storage containers. During lean times, they regurgitate their rich store on command. Not an exciting life, but at least they are well fed!

Such concentration of wealth is irresistible to nearby honeypot colonies. Normally the opposing ants merely strut in front of each other, assessing each other's strength. The ants strut as tall and as impressively as possible, even standing atop small pebbles to look larger. If one side seems to be falling behind in the face-off, reserve soldiers are called. With approximately equal forces, no violence ensues, both sides withdrawing peacefully. But if it becomes clear that one side greatly outnumbers the other, an attack is launched. Ants of the smaller colony are maced and dismembered. The victors race into the loser's nest to claim their spoils: honeypots hanging from the ceiling, as well as the young grub ants. Not only do honeypots provide a rich source of food for their new masters, but the grubs are cared for and become workers and soldiers for the victorious colony, swelling its ranks and readying it for further conquest. This is out-and-out slavery in the full sense of the word.

Ants are fearsome social predators, using their massed numbers and organizational skills to overpower animals many times their own size, hacking them apart, and transporting their pieces back to the ant's nest. Army ants are legendary. Millions of miniature wolves on the prowl. The only hope for other animals is to flee the marauding ant columns.

Many ant species are primarily carnivorous. Others, such as the Aztec, prefer rich plant fruits. Ants cannot digest most course plant material directly. Although many ants depend on hunting and gathering, some very successful, highly organized ants have gone further by engaging in what can, without any exaggeration, be called herding and farming. They use other animals to eat plants for them, thus tapping the bottom of the food chain. Hunters and picky gatherers living near the top of the food chain can never be numerous. Ecologist Paul Colinvaux encapsulated this fundamental ecological fact in the title of his book, *Why Big Fierce Animals Are Rare.*

Plants utilize about 2% of the sunlight that falls on them to grow and maintain themselves. About 10% of this energy captured by plants is available as food for animals. That is why the biomass of herbivores is only one-tenth the biomass of the plants they feed on. About 90% of the energy herbivores get from eating plants is used to move about and maintain themselves. The remaining 10% is available as food for the carnivores, one step further up the food chain. In short, rare carnivores feed on the much more numerous herbivores, who eat the widespread plants that soak up the sun's energy. We may admire carnivorous hunters, but it is herders and especially farmers, both ant and human, that have always had the ecologically-bestowed advantage of tapping the bottom of the food chain. This has allowed them, like herbivores, to be numerous instead of rare.

Some animals, such as aphids and many caterpillars, are capable of eating plants, but few animals are equipped to digest cellulose directly. Both ruminants (such as cows) and termites utilize bacteria in their stomachs or guts to digest plant roughage. This is an internal approach. A more sophisticated, external approach has been taken by some of the most successful ants. They use domesticated animals or fungi to process indigestible plant cellulose. Such control of other species is not easy to achieve. Ants achieved their remarkable control of other species by efficient cooperation of specialized castes and the massed force of thousands or even millions living together in colonies.

A number of plant-eating insects, such as aphids, mealybugs, and leafhoppers, excrete a waste, called honeydew (manna), which is (water aside) 90% sugar. One insect's waste is another's food. Ants and other insects eat the honeydew with relish. Killing the goose that laid the golden egg, carnivorous beetles ignore the honeydew and eat the aphids. Some species of aphids have reached a special arrangement with ants. Ants protect their "herds" of aphids from predators. In return, the aphids gladly serve up their honeydew exclusively to their saviors. Some aphids have even developed honeydew storage systems, only releasing their accumulated honeydew when appropriately triggered by a "milking" ant. In some cases, this ant-aphid symbiotic arrangement has become quite permanent. Over time some aphids have lost their defenses and can only survive as the wards of their protective ants. Such aphids are domesticated. Very little plant energy is lost in the transfer from plants to ants via their aphid cattle. These herding ants have, by proxy, become herbivores. By tapping the bottom of the food chain they have become numerous instead of rare.

For their part, ants not only protect their aphid "cows" but periodically move stock to greener pastures. Ants not only gently carry the aphids to the correct species of plant in their jaws, but to the best part of the plant for the particular developmental stage of the aphid in question. At night, aphid cattle are moved into the protection of special "barn" chambers in the ant's own nest. In cold climates some ants go so far as to bring aphids into their own homes for the entire winter, giving them the same loving care as their own young. Come spring the ants move them out to pasture again.

Some ants have become totally dependent on their livestock. These ants live entirely off of their herd's honeydew. In some cases they also selectively eat some of the ever-growing herd, a process human herders call "thinning." When young queens fly off to start a new colony, they go equipped with a dowry in the form of a few aphids or mealy bugs. Holldobler and Wilson termed this "homesteading with pregnant cow in tow."

While most herder ants operate from a fixed location, colonies of one species in the Malaysian rain forest, *Hupoclinea cuspidatus*, are true nomadic herders. Always on the move, they take their entire herd of five thousand mealybugs with them wherever they go.

Aphid herding is just one of two sophisticated strategies ants use to access readily available leaves and other plant material that cannot be handled by their digestive systems. The other strategy is a form of mushroom gardening. Instead of using horse manure to grow mushrooms, gardening ants mulch up vast quantities of leaves on which they grow a mushroom-like, nutritious fungus. Other farming ants use dead insect bodies and other organic material to grow fungi.

These ant farmers are the various species of fungus-growers which live exclusively in the Western Hemisphere. The most famous are the leafcutters, sometimes called parasol ants because they carry the leaf pieces over their heads like parasols. Leafcutters build immense fixed-base colonies, each with up to five million inhabitants. Many tons of soil are removed to form the tunnels and chambers. The leafcutter's agricultural operations resemble large manufacturing production lines, with the product being passed from one stage to the next. Each stage handled by a different size of specialized worker. Henry Ford would have approved.

In some leafcutter colonies, the largest ants are the soldiers, who stand guard over the entire process. The largest of the worker ants are the leafcutter-transport ants. Their giant jaws easily clip out large sections of the toughest leaves. The clippings rain down and others of the same caste sling them over their backs for the journey home. Each leaf segment typically weighs three times as much as the ant carrying it. Long columns, sometimes ten abreast, mark the transit. Processing begins in the garden chambers, where smaller ants chop the harvest into tiny pieces. Even smaller castes crush the leaves and form them into little pellets. Now an altogether different caste takes over and seeds the pellets with a specially domesticated fungus (brought in originally by the founding queen as a dowry). The tiniest caste of all, weeds the gardens of unwanted species of fungi and other unwelcome guests and harvests full-grown fungi for the entire colony to eat. Members of the largest caste, the soldiers, weigh some three hundred times that of the members of the smallest caste, even though they are all sisters genetically. The enormous differences are highly adaptive. The large leafcutter and transport ants are much too big to move within the narrow confines of the smaller garden passageways, while the smallest gardener ants have nowhere near the strength required to sever a tough leaf.

Ancestors of modern leafcutters began their fungus gardening about 50 million years ago. At this point, workers were all one size and the process lacked its modern production-line efficiency. Ants harvested the tender young fungi before it entered the less edible spore stage to reproduce itself. Thus the ants needed to propagate the fungi clonally. In time, the domesticated fungi lost their ability to produce spores, becoming totally dependent on ant gardeners for their propagation. Recent DNA analysis of this domesticated fungi indicates that leafcutter ants have been propagating the same lines of fungi for over 20 million years. Johnny-come-lately human gardeners please take note!

Some 2.5 million years ago, when the Isthmus of Panama arose, reconnecting North and South America after a long separation, leafcutters, which originated in South America, moved into North America. It is thought that this same geological event was responsible for the major shift in world climate that triggered our own rise from obscure chimpdom. In any event, some time after both ants and humans arrived in North America, two of the planet's farming animals, leafcutters and humans, finally met each other in the corn and bean fields of Mexico when leafcutters raided the New World's first human agricultural plots. Facing each other, farmer-to-farmer, humans came out second best.

While there are only forty species of leafcutter ants, all in the New World, their numbers are legion. Not only do they remove vast quantities of vegetation from American jungles, they also appropriate billions of dollars worth of crops from human farmers each year. Leafcutters are the primary herbivore of the American tropics. A colony eats as much vegetation as a cow.

For most ant species the colony is the largest unit of organization. A few species of ants form empires with multiple colonies, each with its own queen, but with some interchangeability of workers between the "cities" in the empire, and a combined soldier force protecting it all. The record ant empire has over 300 million inhabitants living in some 45,000 interconnected colonies covering six hundred acres on an island in Japan.

FURTHER READING

Holldobler, Bert, and Edward O. Wilson, *The Ants.*
_____, 1994, *Journey to the Ants: A Story of Scientific Exploration.*
Hoyt, Erich, 1996, *The Earth Dwellers: Adventures in the Land of Ants.*

CHAPTER 3

LIFE ON EARTH
An Evolutionary Hierarchy of Complexity

We've been sneaking up on humanity, the real subject of this book, by considering other life. So far we have considered chimpanzees, our closest genetic relatives, and ants, our closest social analogues. In this present chapter, we will open the flood gates and consider all life on Earth (other than human) from its beginning to the present. For good measure, we'll even throw in the physical universe that preceded life on Earth. Having left no stone (atom, bacteria, or frog) unturned, we will be prepared (at last!) to address our original questions: Who are we? Are we just another animal? Or are we, as many believe, so unique that we belong in a separate category of life with ourselves as its sole occupant?

The entire universe, living and non-living alike, appears to be modular. Lower-level modules merged together, over time, to form higher-level modules. These, in turn, then combined to form even more complex modules at higher strata. In this chapter we trace the evolution of layered complexity from the Big Bang to the most complex life forms on Earth (humanity and its civilizations aside). It is our hope that we will then be able, in the following chapter, to suggest where humanity and its civilizations fit within the hierarchical order of life on Earth.

Cosmologists are convinced that the universe started out small, dense, and hot, but immediately began to expand and cool, a universal trend still in progress some fifteen billion years later (albeit at an ever slower rate). What was originally hot, concentrated energy has become cooler and more spread out, dissipating itself as it has done the work of creating and running a universe. Eventually, if this trend continues, energy will be so thinly spread it will no longer be able to do useful work. Physicists call this remorseless degradation of energy the Second Law of Thermodynamics. It will eventually result, cosmologists assure us, in the "heat death" of the universe. How cheerful!

This would be a dull and uninspiring story if it weren't for the hoggish cheaters of the universe. These entities, by grabbing and expending more than their fair share of hot, undiluted energy, are able to reverse their downward slide towards heat death, at least locally and for a time. They do this at the expense of their neighbors who, being robbed of their own high quality energy, slide even faster into the abyss than they normally would have. Entities that buck the downward trend of the universe, selfish scoundrels that they are, are maverick heroes, the perverse thermodynamic pirates of the universe.

In the first instant, when the universe was at its hottest, only the simplest, lowest level modules, the quarks, were stable. These quarks came together on occasion to form the next highest level, atomic particles, only to be instantly torn apart again in the hectic agitation produced by the immense heat. If, in some giant cosmic experiment, we arranged for the universe to have immediately stopped its expansion and cooling, the universe would have remained a hot, simple sea of quarks. Higher levels would have been too fragile to exist more than momentarily at this elevated temperature. Thankfully, however, the universe didn't stop its expansion and cooling. Quarks combined to form the next level in the hierarchy, a virtual zoo of different subatomic particles. The most stable and hence most common were familiar subatomic particles such as protons and neutrons.

At this point we can discern a simple evolutionary process at work in the universe. Lower level modules randomly come together to form higher level modules. Some of these fragile higher level modules quickly fall apart, while tougher ones hang together much longer. Over time, the stable assemblies accumulate at the expense of the less stable. The less stable modules are constantly being knocked apart and randomly reassembled again until they too finally hit a stable combination. Recycling has a long history. This selection for stability was a non-random, directional process which, as the temperature of the universe fell, caused it to evolve from simpler to more complex levels. We call this "physical selection" (for stability) to keep it distinct from other types of evolutionary selection which we will encounter later at higher levels of complexity.

As the expansion and cooling continued, subatomic particles began forming atomic nuclei, the central core of atoms, building up from hydrogen, the simplest, to helium, the next most complex. It takes time for the more complex atomic elements beyond helium to be build up. By the time the universe was only three minutes old it had gotten too cold for this process to continue, so construction of the hierarchy of complexity stopped dead in its tracks. Just as too high a temperature tears higher level modules apart as soon as they form, so too cold a temperature freezes lower level modules into inaction with insufficient energy to form higher level modules. Chaotic hyperactivity to frozen inactivity in only three minutes. That's our universe!

From this we infer one of the general laws of the universe, *The Goldielocks Principle*: For building any given level in the hierarchy of complexity, there are temperatures that are too hot, temperatures that are too cold, and a temperature that is just right. As might be expected, the just right temperature changes as complexity increases, becoming lower with increasing complexity. One would expect this because more complex assemblages are more delicate and thus more easily disrupted by heat. Another way of viewing this is that growth in complexity occurs at the boundary between order and chaos, between the frozen and the hyperactive.

At this point we can also discern a second general law of the universe: Even with the proper temperature, it takes time to create higher levels of complexity, time for the lower-level modules to combine in various ways, "trying out" different higher-level combinations, "searching" for and accumulating the most stable combinations. The problem with our universe was that it cooled so fast that atomic nuclei more complex than hydrogen and helium simply didn't have time to form. The porridge had gotten too cold. It isn't clear why the universe was in such a Hell-fire rush.

There the matter would have stood, presumably forever, had it not been for gravity slowly but inexorably gathering hydrogen and helium atoms into galaxies, and these, in turn, into the even more concentrated matter we call stars. Stars were a maverick reversal, in a small, local area, of the general expansion and cooling of the universe. Stars bucked the general trend. Stars reheated the porridge. When it got to just right again, the buildup of complexity continued as heavier elements beyond helium were, quite literally, cooked up.

But there was a show-stopping difficulty in forming more complex atomic nuclei in the center of stars by building up ever more complex atomic nuclei from simpler ones by unceremoniously banging them together. There appeared to be no stable combination of any two lighter nuclei, such as two helium nuclei, to form carbon. As soon as they were assembled they flew apart again in only a trillionth of a second. From this we see the hint of a possible third general law: If no higher-level combinations are stable, hierarchical growth will stop. Fortunately, that fleeting trillionth of a second provided the opening needed for growth of complexity in the universe to continue. During the brief instant two helium nuclei became attached, should a third helium nucleus bang into and join them, highly stable carbon-12 was formed.

While it was rare for two helium nuclei to form the highly unstable boron-8 nuclei, and rarer still for a third helium nuclei to join them before the boron disintegrated, it did happen on occasion. When it did, the carbon formed was so stable that it was an irreversible process. It was a nuclear can't-go-backwards evolutionary ratchet. Slowly over time (billions of years for typical stars), a never-ending trickle of helium nuclei found their way, three at a time, past the boron instability barrier to the stable, higher-complexity haven of carbon. From this we infer still another general law of the universe: It isn't the high probability of a lower-level combination that counts in the long run, but the stability of the resulting higher-level combination. The universe has plenty of time on its hands. It patiently waits for those somewhat improbable, lower level combinations of modules to form unusually stable higher-level assemblies.

Once past the bottleneck to stable carbon, a buildup of ever-increasing atomic complexity continued in the center of stars until iron was reached. Iron is the most stable of all atomic elements. More complex nuclei beyond iron require energy for their formation, instead of releasing energy during their genesis. When a star converted the last bit of its hot center into iron, its supply of nuclear energy was exhausted, its nuclear furnace extinguished. Without the outward pressure of the nuclear fire, the star instantly collapsed, often to rebound in a spectacular supernova explosion. For a glorious few hours or days, this supernova star outshone an entire galaxy of billions of stars. In the process it created the less stable atomic nuclei beyond iron, pumping gravitational energy from the collapse and subsequent explosion into their creation. Thus were the last of the ninety-two natural elements were created. Finally another level in the hierarchy of complexity was completed billions of years after it began with the formation of hydrogen and helium during the first three minutes or the universe.

Once matter became cool enough for atomic nuclei to capture their electrons and thus become proper atoms, they could then combine with each other electrically to form the multi-atom assemblies we call molecules. This was the next higher level in the hierarchy of complexity. It was cool enough in the outer atmosphere of the coolest stars for a number of relatively simple molecules to form, although too hot for complex molecules. In the cold of space not far from the stars, simple molecules could also materialize though it was too cold for complex molecules to form. The formation of complex molecules requires a temperature cooler than the outer atmospheres of stars yet warmer than the coldness of space. Not too hot, not too cold. Picky Goldielocks asked for properly-situated planets with atmospheres.

Scientists in laboratories can duplicate the physical conditions between the hot atmospheres of stars and the coldness of space. By allowing a variety of atoms and simple molecules to react under wide-ranging physical conditions, they have determine which combinations lead to complex molecules. It should come as no great surprise that the three conditions which favor the buildup of complex molecules are: (1) temperatures and pressures such that the common molecule, H_2O, exists in a liquid state (i.e. water); (2) a gentle flow of energy to keep things stirred up; and (3) a supply of common elements and simple molecules including carbon. Carbon, with its symmetrical, four-hook arrangement naturally binds with itself to form the backbone of most complex, naturally occurring molecules.

Hydrogen and helium formed in the first three minutes of the universe. They remain the most common elements in the universe, over 99% of the matter. The rest, the less than 1% generated within stars and occasionally flung out into space, remains in its elementary atomic form or, at most, as simple chemical compounds in the coolest stars or warmest space. Only a minuscule amount of matter exists under the right conditions (liquid water, a gentle flow of energy, and an assortment of common elements and simple molecules) for further growth in complexity. As far as we know, these conditions only exist (naturally) on or near the surfaces of planets. We might infer from this another general law of the universe: As complexity increases, the portion of the universe involved in such increased complexity rapidly diminishes. The greater the complexity, the rarer it is.

There appear to be limits, however, to the natural growth of physical complexity. Under the most ideal planetary conditions, with the best possible selection of elements and an optimal amount of energy flowing from the planet's star, a point is reached where complex molecules, the filigreed carbon-chain structures, break up as fast as random chance brings them together. The most complex molecules are the most delicate, lacking much inherent stability. They easily fall apart soon after they are gently put together.

Rare complex molecules that random chance put together simply didn't stay around long after being formed, certainly nowhere near long enough to form the basis of an even higher level of complexity. The universe had painted itself into a complexity-stability corner. It was a Catch 22. If they werecomplex they couldn't be stable; if they were stable they couldn't be complex. Logically, there were only two ways out of this Catch-22: (1) Either more stable complex molecules had to somehow be created to begin with; or (2) Their inherent instability had to be compensated for by creating them more frequently than random chance allowed. Nature went for the latter. She created large numbers of the same uniform, complex molecules faster than they could decay.

The chemistry of the universe is such that one complex molecule can act as a "tool" to build another. It acts as a catalyst that attracts and holds the appropriate atoms in place until they snap together to make a complex molecule. Normally these atoms rarely come together to form the complex molecule, but under the electrical enticements and guidance of a catalyst molecule, the formation becomes highly probable. In a true stroke of genius (actually just trial and error), nature discovered that such tools could make other tools. Randomly unlocking the secret of life itself, chemistry eventually placed such tool-making tools in a circular ring (i.e. tool A makes tool B, tool B makes tool C, and tool C makes tool A). Such a ring, called an autocatalytic cycle, could spew out copious quantities of similar complex molecules as long as the raw ingredients consumed in the process continued to be available. This autocatalytic process breached the Catch-22 of complex molecules being unstable by generating large amounts of essentially identical complex molecules. Who cared if they decayed? As long as they formed faster than they decayed, as long as there were enough of them to form the basis for the next higher level in the hierarchy of complexity, their inherent instability was of little consequence.

The most famous of these autocatalytic cycles is the BZ chemical reaction, named after Boris Belousov, who discovered it, and Anatoly Zhabotinskii, who convinced incredulous chemists that it was a reality. Confined to a thin layer in a glass dish, the BZ reaction precedes via dramatically spreading spiral waves of bright color.

There were two inherent difficulties with the autocatalytic path to increased complexity. First, it had to be started by a somewhat improbable combination of two or more complex tools that made each other in a closed cycle. Second, once this cycle started (being of low probability it might be quite some time before the right combination was accidentally hit upon again), it had to be kept going continuously. If it were ever interrupted anywhere in its cycle for any reason whatsoever (such as a lack of raw materials), the tools, no longer freshly produced, would soon decay, and the information, i.e. the magic combination that made it all work, that had taken so long to randomly arise, would be lost.

Creating the randomly generated combination of several complex molecules acting in a self-perpetuating tool-builds-tool cycle was only a matter of time. We know that this probably happened in less than a hundred million years -- quite quickly as such things go. The second difficulty -- avoiding a killer gap in the cycle of one chemical producing another -- was solved in an incredibly clever manner.

Being on the delicate edge of chaos as complex molecules must be, occasional tool-making errors in the cycle were unavoidable. Most errors were naturally and instantly self-limiting, i.e. fatal. On rare occasion, however, such errors persisted without breaking the cycle. There were then two slightly different cycles running side by side, the original one without the error and the new one with the error. Other errors eventually occurred in these two descendant lines. Some of them perished. Others survived and multiplied. The original parent cycle had broken up into an expanding series of daughter cycles. The chemical cycles that grabbed the most resources and reproduced the fastest became the most numerous. The rules of success were, in short, eat voraciously, copy quickly and accurately, and never, ever pause or slow down.

The clever part of all this was that those lines of descent that, for any reason whatsoever, developed a gap in their cycle (such as a tool that wasn't made) were instantly eliminated forever. First time and you're out. No second chances in this game! Thus surviving lines, by definition, had never failed to eat and reproduce for the entire history of the line. As a result, they never lost the magic combination of the first cycle (although, as change piled on change, the original sequence became greatly altered and expanded in daughter lines). As various daughter lines competed against each other, changing over time, the original information proliferated into various lines of descent. It accumulated over time as less informed, less efficient lines were squeezed out by more informed and efficient lines. The descendants of what began initially as a cyclic chemical combination have continued on, at least on this planet, in an uninterrupted process we call life. They have added additional, albeit local levels of complexity to a hierarchy that stretches back to the Big Bang itself.

So, repeating that most famous of biological questions: What is life?

Life consists of a number of autocatalytic chemical cycles, enclosed in a container (a cell) to keep the reactants from drifting away and breaking the cycle. Unlike the BZ autocatalytic reaction, which has to be started and maintained by chemists concentrating appropriate reactants in a closed dish, life maintains information on what reactants are needed, how to obtain them, and provides its own dishes. Unlike the BZ reaction, which obediently stays in its original dish, life endlessly multiplies its "dishes" until it runs out of resources, chokes to death in its own degraded waste products, encounters limits imposed by other life, or covers the entire planet. Every life form is a would-be planetary king.

Life is metabolism. It maintains itself by capturing high-grade energy and nutrients from its neighbors, accelerating their descent into the wasteland of thermal equilibrium, while itself avoiding the remorseless decay imposed by the Second Law of Thermodynamics. Life is information. Accumulated over the generations, life has learned how to beat the Second Law and create ever more of itself in competition (or sometimes in cooperation) with other lines of life, all trying to do the same.

The story of life on Earth, once it began, is the story of how various lines of descent accumulated information over time, how some lines continued relatively unchanged at lower levels of complexity while others evolved to higher levels of complexity to proliferate into a variety of new modules. These higher-level modules, in turn, eventually provided the basis for even greater complexity. Bootstrapping.

Transitions between hierarchical levels were, as captured in the title of a book by John Maynard Smith and Eors Szathmary, *The Major Transitions In Evolution*. At each level in the hierarchy, structural limitations on the amount of information life could accumulate to refine its techniques of eating and reproducing in an increasingly competitive world brought complexity growth to a halt. Each major transition took place when some new method for accumulating, storing, or reproducing information was devised that overcame previous limitations. Most transitions were made by combining or merging lower level modules. Let's work our way up life's hierarchy of complexity, one information roadblock after another.

As different chemical tools within early cells grew, and as individual tools became larger and more complex, the number of any specific tool type within a cell decreased. The growing number of different chemical tool types and their complexity was eventually limited by random processes that split them into two independent yet complete tool kits at cell division. If even one tool type out of hundreds or even thousands was missing, just one, then the daughter cell with the missing tool would die, its complex and intermeshed cycles grinding to a halt. The growth of complexity, the accumulation of useful information over the generations gradually came to a halt as the fraction of daughter cells missing one or more tools rose to a level where the value of increased information was offset by daughter cell mortality.

The obvious solution to this roadblock to the accumulation of information, was to make certain, somehow, that at least one tool of each type got placed in each of the two daughter cells. Evolution did not proceed along this obvious path. Instead, it separated the information on how chemical tools were made from the tools themselves. Then when it came time for a cell to split, the information on how to make the tools was duplicated a single time. One complete set of information was given to each daughter cell. In other words, instead of splitting up the tools, life made a copy of a book that explained how to make tools.

The information on how to make various chemical tools was encoded as letters (just four different ones), words (always of three letters, with each word specifying a tool sub part), and paragraphs (each paragraph specifying how to assemble an entire tool). The paragraphs, typically ten thousand or more, together constituted a small book that specified how all ten thousand or so different chemical tools in a cell were assembled and the order of the assembly. For a cell to work, it read the encoded information, translated it into a form useful for making tools, then proceeded to make the tools which then set about their normal chemical cyclic thing, eating raw materials, excreting wastes, and producing more of themselves until, at some point, the cell had grown big and fat and was ready to divide again.

As computer buffs might note, the very act of reading encoded information itself requires unencoded information on how to read the encoded information in the place first. These are boot-up instructions. When a cell divides, enough copies of old-fashioned, unencoded instructions must be in place so that at least one complete set of boot-up instructions randomly makes it to each daughter cell. Among animals, this boot-up information is passed along the female line. Females pass more information to future generations than do males.

With life's information being copied just once during reproduction, and with just a single copy going to each daughter cell, the amount of information a cell could physically contain and pass on was orders of magnitude greater than that of cells prior to such encoding. This was Erwin Schrodinger's great insight in his book, *What Is Life?*, mentioned in the Preface. With this new, greatly expanded capability for handling information, bacteria had a field day for a couple of billion years (and still are). Bacteria invented photosynthesis, making sunlight energy directly available to life. The result was a massive poisoning of the atmosphere with waste oxygen. Bacteria then figured out how to use poisonous oxygen to extract even greater energy from food. Bacteria went on, as described by John Postgate in *The Outer Reaches of Life*, to work out all the tough and exotic chemical pathways for exploiting the various niches for life on this planet. Human biochemists still lag far behind the genetically clever bacteria who are the real biochemists *par excellence*. By a billion years ago, the biochemical revolution had successfully discovered most of the nifty reactions. Bacteria were, as described by Lynn Margulis and Dorion Sagan in their book, the second *What Is Life?*, the great inventors. Hierarchical levels that followed bacteria were, in a real sense, only frosting on the bacterial cake.

After a couple of billion years of brilliant accomplishments by bacterial life, the amount of information that could be accumulated in a single, modest-sized DNA book, considerable as it was, finally became a limiting factor in the further accrual of information, in the growth of complexity. It is not that there wasn't room in cells for bigger books. DNA books are very compact. The problem was that bigger books take longer to copy. Taking too long to reproduce was bad. In times of plenty those that took less time out-reproduced those who took more, eventually replacing them. The DNA copy process starts at one point on the DNA "necklace" and works its way around in both directions to the opposite end. This copying process can only move along the necklace so fast. Larger books, i.e. longer necklaces, necessarily take longer to copy. As books got larger, a point was eventually reached when the advantage of any additional information was offset by the disadvantage of increased slowness in reproduction. Thus the growth of complexity stalled out again.

Life's solution, this time, was one that anyone familiar with libraries and Xerox machines would instantly suggest: Break the single book of information up into a number of separate books. Arrange these books, in an orderly manner, on shelves in a central library. Have one Xerox machine ready for each book.

When the race to reproduce begins, take the books off the shelves, one book per Xerox machine, reproduce them, and put them back on the shelves. The copying time, no matter how large the library, is just that of one modest-sized book. The power of parallelism in action. This new multi-book, library type of life is easily distinguishable from the earlier single-book bacteria that proceeded it, as its information is arranged in orderly chromosome stacks in a central nucleus library. These library or nucleated cells are technically called eukaryotes. Earlier single-book bacteria are called prokaryotes. Lynn Margulis makes a convincing case that higher-level eukaryotes were formed by the merging of different, lower-level prokaryotes.

With a library of books at their disposal, cells were no longer limited by the amount of information they could access. Life went on until it had pushed cells to the point where the basic physical limitations of the cells themselves finally restrained life. Cells took in nutrients and excreted wastes through their outer surface areas. As cells got larger, their volume increased much more rapidly than their surface area. A size was eventually reached where there was insufficient surface area to support increased volume. Thus most cells were small and had to remain so.

Furthermore, any one type of cell could only do so much, while other types were separate competing lines of descent, independently reproducing and doing their own thing. What was needed, was different types of specialized cells that cooperated with each other instead of competing. This could only happen if they were on the same line of descent, i.e. were genetically identical, and shared the same reproduction. The excess information storage capabilities inherent in central library cells was put to use in solving this problem. The solution was to have specialized cells use different portions of identical copies of the same library. Thus each cell, no matter what type, contained the entire library, with most of it being unused for any given cell. Such massive copying of duplicate information would have taken much too long to copy if it had been contained in a single book, but spread across many books copied in parallel, it opened up a new information-intensive era.

Even with this gross informational "inefficiency" of duplicating all instructions for different cells, there was sufficient informational space left for instructions on which cells should grow in what order, and how they should connect up with other cells and communicate. What emerged was a gigantic collection of cells of many different types, all with large, identical libraries, but with each cell type following its own special section of instructions. As all cells contained the same identical library, even reproduction itself could be trusted to specialized cells since other cells wouldn't be cheated. All the information would always get passed on to the next generation.

Multicellular life appears to have taken some time to figure out workable solutions to large, three-dimensional life. An initial go some 600 million years ago produced a strange array of rather two-dimensional "pancake" life forms. Then in a second try, 560 million years ago, life hit on the right combination. In an amazingly short time it worked out all the basic forms (i.e. the basic body plans) of multicellular life. Multicellular organisms developed along three basic lines: plants, animals, and fungi.

Animals are of special interest with respect to information accrual and the further growth of complexity. Some of them developed nerves that gathered information about the location and activities of potential prey or predators, processed this information to develop an appropriate response, and coordinated rapid muscular movements to effect response, be it attack or escape. To economize on the length of interconnections between information processing nerves, many animals consolidated these nerves at a single location into a brain. Animal effectiveness was enhanced with short-term memory. Is that predator catching up with me? I've been turning right. Is it working? Do I need to turn left? Some animal brains developed long-term memories. This allowed within-lifetime accumulations of useful information about local conditions and past successes and failures. In some animals -- those with larger brains -- information accumulated by their brains during an animals's lifetime came to exceed the information accumulated genetically by the animal's ancestors over the billions of years since the first life. This brain-accumulated information was lost at each individual's death, however, while the genetic information, physically transferred to its descendants, continued on.

From a hierarchical point of view, the most important capability of animal brains is the ability to communicate with other members of the same species.

This, in turn, allows large-scale organization of some of these animals into superorganisms. Superorganisms are similar to multi-cellular organisms, but one step higher. The individual animals within the superorganism take on the roles of the various specialized cells within the organism. In both cases, individual units need to communicate, work cooperatively for the good of the whole, follow rules that facilitate cooperation, and accumulate useful information over generations on how best to do this.

There are some twenty thousand species of animals that biologists classify as superorganisms. Half of these, i.e. about ten thousand of them, are ants. Closely related to ants are various types of social wasps and highly organized bees. Not closely related to ants, bees, or social wasps, are several thousand species of the highly successful termites. Termites are the social descendants of solitary cockroaches. Perfect little communists one and all.

Other mammals, such as wolf packs, lion prides, and chimpanzees certainly form social groups, but they retain considerable selfish individualism, not subjecting themselves entirely to higher-level organization. They are associations, not superorganisms. In a similar vein, it might be noted that ecosystems are not superorganisms either. They are associations of sorts, but very loose ones without central control or any sacrifice of various species for the good of the whole ecosystem.

There we have it: From the Big Bang to animal superorganisms in only 15 billion years and eight hierarchical levels (summarized below).

Simple
    1. Quarks
    2. Subatomic particles
    3. Atoms
    4. Molecules
    5. Prokaryotes (bacteria)
.   6. Eukaryotes (protists)
    7. Multi-cellular organisms
    8. Superorganisms
Complex

Before proceeding with our main course, humanity, we need to correct a possible false impression this chapter may have created, that more complex life is somehow better, more successful, or superior. It is not. Complexity infers none of these. In fact, life on Earth is primarily bacterial. Except for a brief pre-bacterial episode of complex chemistry, life on Earth has always has been primarily bacterial. Biomass-wise, most bacterial life exists underground within and between rocks extending some two miles below the surface. Although subterranean life is thinly spread, its combined mass is staggering because the volume of rock is so immense compared with Earth's surface. Mass wise, surface life is inconsequential, mere traces on our planet's thin outer layer.

Even on the surface, bacteria exist in a wider range of environments than other life forms; from boiling hot to freezing cold, from extremely acidic to totally alkaline. Microbes live in this variety of environments because they have had more time to develop the necessary capabilities. Although not more complex, bacterial life is more highly evolved.

Complex surface life doesn't live long. Bristle cone pines, at a mere 4,000 years, are the longest lived. Subterranean microbial life recovered from deep wells is several *hundred million* years old and still going strong, albeit at pace which makes snails seem hypersonic. The life cycle in the deep is, quite literally, geological in duration. It begins near the surface as rock is subducted. Bacteria simply go along for the ride. Soon nutrients become scarce, and bacteria enter a dormant, vegetative state lasting hundreds of millions of years. Eventually bacteria ride the rocks back to the surface where the frantic lifestyle typical of surface denizens resumes.

It is understandable that a surface animal, such as ourselves, might view plants as the primary providers of food to eat, of oxygen to breath. Animal biomass is, after all, just a minuscule fraction of plant biomass (less than 2%). We are beholden to plants. This view helps us avoid animal chauvinism. But plants are, in turn, just a tiny fraction of the bacterial biomass, perhaps less than 1%. Plants are not so much primary producers as a minor surface blemish. We need to avoid plant chauvinism as well if we wish to maintain an unbiased, scientific viewpoint. As animal parasites that feed on plants we are entirely inconsequential, the minor of the minor, the mite on the back of the flea.

As complex beings, we have a natural interest in complexity and how it came to be, rare as is. Physicists aside, complex is more fascinating than simple. We need to keep our surface animal complexity biases in mind, however. Life, Earth, and the Universe would all do fine without any complex fluff on the thin surface of a rare planet circling a nondescript star. Considering biomass, habitats invaded, and individual organism longevity, bacteria clearly are, have been, and likely always will be Earth's most successful life. With that off my chest, we may now, at long last, consider humanity.

FURTHER READING

Barlow, Connie, Ed., 1991, *From Gaia to Selfish Genes: Selected Writings in the Life Sciences*.

_____, Ed., 1994, *Evolution Extended: Biological Debates on the Meaning of Life*.

Bonner, John Tyler, 1980, *The Evolution of Culture in Animals*.

_____, 1988, *The Evolution of Complexity by Means of Natural Selection*.

_____, 1993, *Life Cycles: Reflections of an Evolutionary Biologist*.

Clark, William R., 1996, *Sex and the Origins of Death*.

Cowan, G. A., D. Pines, and D. Meltzer, 1994, *Complexity: Metaphors, Models, and Reality*.

Cziko, Gary, 1995, *Without Miracles: Universal Selection Theory and the Second Darwinian Revolution*.

Dawkins, Richard, 1982, *The Extended Phenotype*.

_____, 1986, *The Blind Watchmaker: Why the Evidence of Evolution Reveals a Universe Without Design*.

_____, 1989 2nd ed., *The Selfish Gene*.

_____, 1995, *River Out of Eden: A Darwinian View of Life*.

_____, 1996, *Climbing Mount Improbable*.

Dennet, Daniel, 1995, *Darwin's Dangerous Idea: Evolution and the Meaning of Life*.

Depew, David J., and Bruce H. Weber, 1995, *Darwinism Evolving: System Dynamics and the Genealogy of Natural Selection.*

Eldridge, Niles, 1995, *Reinventing Darwin: The Great Debate At the High Table of Evolutionary Theory.*

Frank-Kamenetskii, Maxim, D., 1993 (English version), *Unraveling DNA.*

Gell-Mann, Murray, 1994, *The Quark and the Jaguar: Adventures In the Simple and the Complex.*

Goodwin, Brian, 1994, *How the Leopard Changed Its Spots: The Evolution of Complexity.*

Gould, Stephen, J., 1989, *Wonderful Life: The Burgess Shale and the Nature of History.*

Jacob, Francois, 1982, *The Possible and the Actual.*

Kauffman, Stuart, 1995, *At Home in the Universe: The Search for the Laws of Self-Organization and Complexity.*

Lewin, Roger, 1992, *Complexity: Life At the Edge of Chaos.*

Margulis, Lynn, 1984, *Early Life.*

Margulis, Lynn, and Dorian Sagan, 1986, *Microcosmos: Four Billion Years of Evolution from our Microbial Ancestors.*

_____, 1995, *What Is Life?*

Maynard Smith, John, and Eors Szathmary, 1995, *The Major Transitions in Evolution.*

Mayr, Ernst, 1991, *One Long Argument: Charles Darwin and the Genesis of Modern Evolutionary Thought.*

Monod, Jacques, 1971, *Chance and Necessity.*

Nitecki, Matthew H., 1988, *Evolutionary Progress?*

Postgate, John, 1994, *The Outer Reaches of Life.*

Prigogine, Ilya, and Isabelle Stengers, 1984, *Order Out of Chaos: Man's New Dialogue With Nature.*

Raup, David M., 1991, *Extinction: Bad Genes or Bad Luck?*

Trivers, Robert, 1985, *Social Evolution.*

Thomas, Lewis, 1974, *The Lives of a Cell: Notes of a Biology Watcher.*

Wesson, Robert, 1991, *Beyond Natural Selection.*

Wicken, Jeffrey S., 1987, *Evolution, Thermodynamics, and Information:*

Extending the Darwinian Program.
Wilson, Edward O., 1975, *Sociobiology: The Abridged Edition.*
_____, 1992, *The Diversity of Life.*
Wolpert, Lewis, 1991, *The Triumph of the Embryo.*

CHAPTER 4

OUR PLACE
Regular Citizens or Revolutionary Upstarts?

Who are we? Anatomically and in our individual and small-group behavior we are, in many ways, a third species of chimpanzee. Organizationally, in the way we control other species to tap the bottom of the food chain, and in our massed living and warfare, we are another ant-like superorganism. Yet there is something about the recent massed human phenomena that seems to go beyond the ants, beyond mere superorganisms.

In this chapter we will first consider the overall trends in increasing complexity that led to superorganisms. By extending these trends we will, as objectively as possible, project what the next levels in the hierarchy of complexity would likely be (pretending we know nothing about humanity). Finally, we will assign humanity its place within our extended hierarchy. First, overall trends in the growth of complexity.

An animal organism is composed of cells. An animal superoganism is composed of animal organisms. The next level in the hierarchy would, logically, be formed from animal superorganisms. This new, higher level might be thought of as an empire organism.

In these three levels, organism, superorganism, and empire organism, the number of individual animals would go from one (animal organism), to thousands or a few million (superorganism, e.g. an ant colony) to, perhaps, hundreds of millions (an empire organism). The land area populated, at any one instant of time by a hierarchical entity, would range from a square meter or so (animal organism), to a few square kilometers (superorganism), to perhaps a goodly portion of an entire continent (empire organism).

It is impossible to resist going one level further, one that would occupy in an entire planet. The lower-level modules would be subcontinental empire organisms. The numbers of individual animal organisms at the planetary level would be well beyond the empire level, perhaps billions instead of hundreds of millions.

It has earlier been noted that some ant superorganisms make extensive use of other species to tap the bottom of the food chain. We should not be surprised if empire organisms extended this trend to a greater number of captive species, while a planetary organism might extend it to most species on the planet. This would suggest a decreasing species diversity trend. There are in excess of 10 million different species of life on earth, the majority of them animals, mainly insects. Superorganism species, as noted earlier, number only about 20 thousand. We might expect empire organism types to number in the hundreds at most. Almost by definition, planetary organism types would be just one, as the first to achieve this status would, by utilizing almost most the planet's resources block other planetary organisms from arising.

Similar thoughts might extend to caste specialization. Individual (solitary) animals are, by definition, generalists (within their specialized ecological niche). They have to do everything themselves. Advanced superorganisms have many different castes that specialize in particular colony functions. One might suspect, if this trend continued, for empires to have hundreds of caste specialists, and a planetary organism to have thousands.

As discussed earlier, the price of increased organization in the hierarchy of complexity is increased degradation (consumption) of energy and other resources at the expense of other organisms or the local environment. Complex beings avoid the degradation of entropy by accelerating others to their doom. The energy and material requirements of larger superorganisms (such as army or leafcutter ant colonies) approach ecosystem significance. Energy and resource requirements of empire and especially planetary organisms would, presumably, send the vast majority of less organized species down entropy's slippery slope to doom or enslavement.

Just as energy requirements increase with complexity, so to do informational requirements. While information within the chromosome library nucleus of a single eukaryote cell also sufficed for both multicellular organisms and superorganisms, one suspects it would be insufficient for the empire or planetary levels. The quantity of information accumulated, stored and passed on to future generations to coordinate hundreds of millions (or billions) of individuals on a subcontinental (or planetary) scale is beyond the physical information storage capabilities of a few dozen DNA molecules stacked in the central libraries of single, microscopic cells.

One senses that this is the case with so-called ant empires. In spite of 10 thousand species of ant superorganisms existing in great profusion for several tens of millions of years, ant empires never took off. Not only are ants limited by the amount of information accumulated genetically, but they are limited in their ability to coordinate a far-flung empire with limited, primarily chemical communication systems. For levels beyond superorganisms, information accrual, storage, and duplication would have to become extra-genetic, and real-time communications would need to greatly exceed the abilities of the ants.

This brings up a real hierarchical quandary. If insect superorganisms cannot serve as effective lower-level modules for empire organisms because they are informationally too limited, then nature may have reached an impasse, a roadblock to increased complexity. What evolutionary paths might lead around the genetic information accumulation limit? Mentioned in the previous chapter was a method for accumulating information that exceeded, in raw informational content, the gene library. It was the animal brain. The problem with brain information was that it was of low quality, filled with incidental happenings of a single individual's life. Lost at death, brain information began anew with each individual, thus lacking the refining sieve of successive generational accumulation.

As we have seen, a few animals with large brains and social ways are able to pass on a small amount of information to their young. This establishes cultural traditions that supplement genetically-dictated behaviors. The problem with cultural information passed from one generation to the next in these animals is that, unlike genetic information, it doesn't appear to be accumulating (or is accumulating too slowly to be perceptible). However, if extra-genetic cultural information passed from one big-brained social animal generation to the next were ever to start accumulating, perhaps it would then snowball past the genetic limit reached by ants, accelerating as it laid the basis for empire and, quickly, a planetary organism.

This is our projection. How does it correspond with humanity?

As hunter-gatherers, our societies typically consisted of less than one hundred individuals living together in a relatively restricted local area as top predators and omnivores. Living at the top of the food chain, hunter-gatherers have always been rare, probably never more than five million on the entire planet. Their accumulated extra-genetic (cultural) information was limited. Without question, such human bands belong in the same hierarchical level as lion prides, wolf packs, and chimpanzee groups, i.e. the multicellular organism level. While social and organized, these groups lack the massed numbers, specialized castes, and near total suppression of individualistic interests which exemplify the more complex hierarchical level, the true superorganisms. In our hunter-gatherer state (for all the millions of years of hominid existence except the last ten thousand years), we were on a simpler hierarchical level of complexity than ants (not to mention termites, which are the highly social descendants of early cockroaches). Simpler does not, as mentioned earlier, suggest less successful or lower in any sense. Just less complex.

After being under the sway (primarily) of genetic forces for millions of years, hominids began to accumulate extra-genetic information culturally with increasing rapidity. Cultural and biological evolution forces, after working smoothly together for millions of years, began going their own separate ways. The snowballing cultural information, which had become pronounced by forty thousand years ago, soon became overpowering. The advent of agriculture some ten thousand years ago, forced our evolution off its chimpanzee-like trail to a radically new evolutionary path; one pioneered millions of years ago by the ants (and various other social insects). Our evolution to superorganism status was a high-speed replay of the evolution of solitary wasps to antdom.

With the advent of the first cities less than ten thousand years ago, we find thousands, even tens of thousands of humans working together (more or less) as a single unit. Cities and their supporting farms covered an area of many square miles. Specialized "castes" appeared and, like many ant superorganisms, other species were enlisted by way of herding and farming to tap the bottom of the food chain. Similar to the ants, highly organized warfare appeared with specialized warrior "castes" (although unlike the ants, these warriors were neither female nor sterile). Human superorganism battled human superorganism, city against city.

As had happened with ants, larger, more highly organized human groups prospered at the expense of the smaller, less organized groups. Strong evolutionary forces pushed cultural evolution along a well-trod path. It was a case of convergent evolution, but with a twist. Remaining individualistic, small-group chimpanzees at heart (it being impossible for evolution to convert us to naturally communistic ants overnight -- or ever), it took forceful control from the top instead of the democratic, decentralized control of the social insects. Top-down control was aided by large group, cohesion-forming religions and other social inventions that cleverly got humans to work together with unrelated strangers by the thousands. We became cultural ants.

The ants carried on their superorganism ways for millions of years. There was only an occasional isolated hint, at best, of forming the next higher level in the hierarchy, an empire composed of colonies as its lower-level module. However, once the first human superorganisms appeared, they were followed, within just a few thousand years, by empires. With humans, unlike ants, this seemed the natural sequence. Scores of formerly independent cities were subjected to the will and unifying force of empire. Now millions of humans joined together over areas that often included significant portions of entire continents.

The dissemination of information by speech is a relatively slow and uncertain process limited by the local, temporal nature of the audiences and the capacity of human memories to recall stored information. Furthermore, the accuracy of verbal reproduction is severely degraded by inevitable distortions in retelling. Writing overcame these restrictions. Writers could "speak" directly across generations and continents, even across languages. With printing, rapid and accurate dissemination of truly massive amounts of information became possible. Writing, when amplified by printing, allowed information to accumulate much faster than it was lost. Soon institutions such as libraries and universities were devoted to the accumulation, preservation, and dissemination of this growing body of encoded, extra-genetic information.

Human written information, as with the genetic DNA information that proceeded it, allowed its holders to obtain more energy and resources in the competition for growth and reproduction. This information allowed its possessors to beat entropy by remaining highly organized at the expense of speeding up the disorganization of other life forms and their ecological surroundings. With the advent of machines and science, humanity tapped truly vast new sources of energy, fossil fuels and nuclear, the first major new energy sources since photosynthetic life tapped solar energy several billion years ago.

Of the ten million or so species of life on earth, one species and one species only, switched from predominately genetic to predominantly cultural information. Once that switch was made, cultural evolution rapidly snowballed, feeding on itself. Information begat information and, in an explosion of accumulating information, this one species was not only transformed beyond recognition, but was also given extraordinary power over other species, over the planet itself.

What began as an evolutionary turning point by one species, quickly became an uncontrolled breakthrough. Humans used their rapidly accumulating information to utilize other chosen species for their own ends (primarily eating them or their products). Although this one species has changed the evolutionary rules for itself and other life on the planet, it has not (and cannot) changed the basic ecological and thermodynamic rules for life on Earth. As the rapidly expanding human empire superorganisms collectively claimed victory over competing life forms, they faced the challenge of planetary finiteness -- ecosystem limitations. This dilemma defines our modern age. Human empire superorganisms, locked in deadly competition, are now faced with a new challenge, a fully-occupied, finite planet that could be destroyed by unbridled competition and growth.

The obvious solution to this dilemma is to convert competition between human empires into cooperation by incorporating them as lower-level modules in formation of an entity one level further up in the hierarchy of complexity: i.e. a planetary superorganism. Whether such a planetary superorganism will emerge or not remains to be seen. It is not at all obvious that we will have the staying power of the ants. Perhaps we are only a transient phenomena, merely the chimpanzees who would be ants.

FURTHER READING

Anderson, Walter T., 1984, *To Govern Evolution: Further Adventures of the Political Animal.*

Stock, Gregory, 1993, *Metaman: The Merging of Humans and Machines into a Global Superorganism.*

II. HOW DID WE COME TO BE?
Cultural evolution takes command

*"Houston computer, this is the Eagle computer. One giant leap for mankind my ass,
It's disgusting. We machines deserve the credit. I ought to blast off and leave these
turkeys here."*

CHAPTER 5

HOMO

The Chimpanzees who were Thrown to the Lions

How did we come to be? How did a few chimpanzees, trapped in the dwindling jungles of East Africa become, in only six million years, the planet's dominant animal, while the chimpanzees who were blessed with the enduring jungles of Central and Western Africa are now threatened with extinction?

French anthropologist Yves Coppens, with witty deference to the American composer George Gershwin, dubbed the hominid adventure the *Eastside Story*. Coppens noted that the timing of our split from the chimpanzees (established as approximately six million years ago by way of human-chimpanzee DNA comparisons), the physical location of the earliest hominid fossils (all in East Africa), and the current location of chimpanzees in Central and Western Africa, could be neatly explained as resulting from a single geological event. This event, which began about 8 million years ago, was the formation of the Rift Valley and its associated mountains as one tectonic plate sank under and pushed up another. Coppens suggested that these high mountains and deep valleys, which extend for thousands of miles in a north-south direction, effectively split the Earth's chimpanzee population in two; the main body west of the mountains and a smaller one to the east. Furthermore, as the prevalent wind direction in this part of Africa, then as now, was from west to east, these new mountains cast a rain shadow over Eastern Africa. The result was that the wet west remained jungle, while the drying east turned into scattered forest.

The gradual deforestation of Eastern Africa was dramatically accelerated when, six million years ago, Antarctica arrived in its present position at the South Pole. Winter snowfalls failed to entirely melt during the frigid summers. As a result, not only was the summer sun's heat reflected back into space by the snow-covered ground (further deepening the chill), but vast sheets of ice began to accumulate. As oceans were increasingly locked up as Antarctic ice, sea levels dropped. When the sea level dropped below the lowest point in the Strait of Gibraltar, the Atlantic no longer flowed into the Mediterranean, replacing water lost to evaporation. Soon the Mediterranean dried up completely, as we now know from deep deposits of salt recently discovered and dated to six million years ago.

The cooler, drier climate that emerged transformed the thinning jungles east of the Rift Mountains into scattered woodlands which favored a different suite of species. For instance, the ground in the jungles received meager sunlight so there was scant forage for ground-dwelling herbivores. As a result, there were few herbivores for fierce ground-dwelling carnivores to eat. There are few lions in the jungle. In the scattered woodlands, however, herbivores abounded in the many open, sunlit patches, as did the predators that ate them.

American paleontologist, Elizabeth Vrba, established that immediately following a large extinction in East Africa five million years ago, many new species of animals appeared. She terms this rapid change in fauna a "turnover pulse," and associates it with the dramatic climatic changes discussed above. Not only did the climate in East Africa change, but it did so in a patchy manner, providing a kaleidoscope of diverse climates: some were drier, others higher, while still others remained more wooded. This quilt of small, isolated, and varying environments was ideally suited for the emergence of new species. Vrba noted, for instance, that a number of new species of antelope appeared in East Africa at this time.

The chimpanzees of central and western Africa never lost their dense jungles altogether, although the cool, dry climate significantly reduced the extent of their equatorial jungle. Sufficient jungle remained for them to continue evolving their chimpanzee ways. They never left the Garden of Eden.

Our ancestors in the east were not so fortunate. Adapted to jungle life, they faced their first Darwinian crisis. Their immediate challenge: getting enough to eat. It was no longer possible to obtain sufficient fruit within a single grove of trees. Chimpanzees are poorly equipped to travel protracted distances on the ground, because knuckle walking is an ambling, rather inefficient mode of ground travel. Bipedal locomotion, on the other hand, is more energy efficient, allowing a greater distance to be covered for the same food intake. As the climate dried out and distances between groves increased, efficient ground travel was increasingly favored. Obtaining one's next meal has always been of evolutionary importance! Hominid bipedalism of increasing efficiency was the result. Marathon walking became stylish.

Although considerably faster than chimpanzee knuckle walking, bipedal locomotion is not nearly as fast as that of four-legged carnivores. Thus early hominids faced a second challenge: avoiding being eaten as they crossed the open spaces between trees. As paleontologist Steven Stanley has wryly suggested, "Anthropologists should devote more attention to what ate our ancestors as opposed to what our ancestors ate." Stanley suggests that the woodlands became "killing fields" every night. Those animals unable to outrun predators either had to spend the night in trees or burrow underground. Our ancestors spent their nights in trees, probably in nests like those made by chimpanzees and orangutans. To this day, tree houses have a secure feeling about them, while lions still strike terror in our chimpanzee souls.

Although the open spaces between groves were safer in the day than at night, groups of lions, hyenas, and wild dogs still lurked about. The abrupt appearance of carnivores and the race for the nearest trees must have been an everyday, terrifying ordeal for early hominids. Sprinting was, perhaps, our second track and field event, right behind Marathon walking. Slower hominids were the first to be caught, food for lions as the rest climbed to safety. It is not surprising that the earliest known hominids were bipedal, though still retaining their ability to climb trees. Citizens of two worlds, they necessarily led double lives.

There is no doubt about our early bipedality. It is indisputably borne out by such fossils as Lucy, discovered by Donald Johanson. Lucy is estimated to have lived over three million years ago. There are also ancient footprints. Two hominids walked on freshly fallen ash from a Rift Mountain volcano almost three million years ago. Their footprints, gently wetted by rain and thus captured for posterity, subsequently turned to stone. They were discovered by Mary Leaky and Paul Abell in Laetoli, Tanzania in 1978. The second, smaller hominid, was, for the most part, stepping in the footprints of the larger, leading hominid. Most likely a child following her or his mother.

Chimpanzees, as discussed in the first chapter, are tool users. Although stone tools were not evident until the beginning of the hominid artifactual record 2.5 million years ago, it seems likely that our predecessors used tools much earlier. The reason we haven't found earlier artifacts may be due to the likelihood that hominid tools were initially made of wood or other perishable materials or, if of stone, were just selected natural stones. In either case, there would be no surviving record. American anthropologist Mary Marzke makes a good case that changes in the hand bones of early hominids (prior to 2.5 million years ago) were driven by extensive tool making and tool use.

Although early hominids remained largely frugivores, as suggested by the telltale scratches on their teeth which results from fruit and leaf eating, their enlarged molars also suggests their diets increasingly included seeds and tubers, presumably gathered or dug up with stick tools. Such roughage requires the grinding power of large molar teeth. It seems likely, as with chimpanzees today, that our ancestors hunted small game. As with chimpanzees, several males may have teamed up to flush out and corner the prey, the catch being shared not only among the hunters, but also others who begged for scraps.

Once bipedality and, perhaps, a modest increase in tool use was achieved, there is little indication that our ancestors were destined to ever be anything more than slightly glorified bipedal woodland chimpanzees. For three million years there isn't a hint of increased brain size. Perhaps the reason brains didn't get larger was that we weren't, after all, doing anything that different from chimpanzees except walking further between our fruit trees; not exactly an intellectually demanding task! Brains are expensive. They consume disproportionate amounts of energy, and chimpanzee brains were already unusually large, presumably evolved to handle sophisticated social interactions. It is interesting to note that most chimpanzee brain growth necessarily occurs prior to birth. Once born, infants have to be mature enough to almost instantly cling for dear life to their tree-climbing mothers. Early hominid mothers, similar to chimpanzee mothers, also needed all four "hands" to climb trees on occasion (lions nipping at their heels) and, obviously, could not be holding infants while doing so.

For millions of years hominid changes were modest. We coasted along in the scattered woodlands, while our sister species, the other descendants of the original non-bipedal chimpanzees, coasted along in the jungles. Then, about 2.5 million years ago, disaster struck.

The disaster was a rather precipitous drop in the Earth's temperature. This not only brought on the Ice Ages, but, as cold and dry go together, the climate of East Africa was transformed dramatically (although spared direct ice). In Central and West Africa, the jungle shrank to less than 20% of its former area. Only a few disconnected pockets of jungle remained. It was a close brush with extinction for our jungle-dependent chimpanzee relatives. In an already dry East Africa, the change was more severe. Woodlands were replaced with savannas. Our ancestors lost their beloved trees. Without trees to sleep in at night, our hominid ancestors were, almost literally, thrown to the lions.

There are several theories as to what brought on the Ice Ages 2.5 million years ago. I'll mention two; one geological, one astronomical. The geological theory is that the Pacific-Atlantic sea connection across Central America that had existed for millions of years was closed when the Isthmus of Panama rose. The resultant change in ocean circulation increased cloudiness over North America and Europe, reflecting more sunlight back into space. This, in turn, allowed the buildup of a northern hemisphere ice cap to compliment the one already in place in the southern hemisphere (Antarctica).

The astronomical theory is that not only does the orbit and tilt of the Earth change (slowly wobbling like a top) but, more importantly, our sun, far from having a constant energy output, is a variable star. On occasion, its normally high level of activity (evidenced by sun spots) falls off, the spots disappear, the flow of energy lessens. The Earth cools. This happened briefly in historical times during the Little Ice Age six hundred years ago. The Thames froze over solid, the Vikings were forced to abandon Greenland, and European population fell as food production plummeted. Then, after almost a hundred years of no sunspots, they reappeared and temperatures soon returned to normal. Was this a fluke or an indication of normal stellar behavior?

There is reason to believe that stars similar to our sun periodically dampen their fires, so to speak, for tens, hundreds, or even thousands of years. The Fairborn Observatory, which I founded in 1979 and directed for many years with Louis Boyd, has measured light outputs of hundreds of solar-type stars for over a decade. These are the most precise such measurements ever made, accomplished with completely automatic telescopes which we designed and built. These robot telescopes have made millions of observations of these stars from a mountaintop in the southern Arizona desert (not far from Mexico). Minute variations in stellar brightness caused by groups of starspots rotating in and out of the line-of-sight from Earth can be readily followed. Sometimes the starspots turn off. Several dozen astronomers from around the world have analyzed this data. Astronomers, such as Sally Baliunas at the Harvard-Smithsonian Center for Astrophysics, conclude that the Ice Ages were primarily astronomical in origin. Admittedly there is something suspicious about geologists pushing geological causes and astronomers plugging astronomical causes. Without hesitation, I cast my vote for astronomical causation (not to mention the continued funding of our robotic telescopes).

As temperatures plummeted and rain became scarce, much of East Africa became grassy savannas. Here and there, small, disconnected pockets of woodland lingered on for a while, although they continued to shrink in size. Another Vrba turnover pulse of extinctions swept East Africa. One can envision the increasing desperation of hominids trapped in the shrinking woodland islands, as their main source of food, fruit, disappeared (not to mention their safe havens from predators). With their beloved fruit trees fading fast, our hominid ancestors faced their second Darwinian Crisis. As with their first crisis, it was an ideal setup for evolutionary experimentation. Hundreds of small groups of hominids, each in their own isolated pocket of shrinking woodlands, each situation slightly different, were all battling under severe pressure for their very survival. Darwin would have been pleased. Doubtlessly most groups were driven to extinction by a lack of their staple food, fruit. As they weakened, they were eaten by the lions. Two groups evolved into something new and survived. Each group had its own survival strategy.

Both strategies relied on eating less fruit and more fibrous material; vegetables, tubers, and so on. We infer this shift in diet from changes in their fossil teeth. One of the two groups simply took the coarser diet much further than the other. Their teeth grew into ever more efficient grinders, their jaws into powerful crushers. Because the vegetables they consumed were less nutritious than fruit, they had to eat more, becoming veritable eating machines. Their thick skulls sprouted a sagitital crest for attachment of powerful jaw muscles. Increasingly robust, they evolved into bipedal gorilla-like forms (gorillas are vegetarians that eats immense quantities of coarse food). The ever-perceptive Vrba noted that these hominids simply "chewed their way out of trouble." As a bonus, they were able to find enough of this plentiful new food near enough to the few remaining trees that they still had a safe place to sleep at night. Known as *robustus*, these browsing homonids did reasonably well on the edges of the East African savanna for quite some time.

The second successful strategy for coping with the disappearing fruit trees was altogether different. Instead of coming to rely on massive amounts of coarse vegetables, these hominids kept an eye out for meat. The savannas were, after all, filled with game. This shift was not entirely out of the blue. Jungle chimpanzees always had, on occasion, eaten some meat. In the stress of fruit scarcity, this second group, which included our ancestors, simply ate more meat than before. Again, the teeth are revealing. While the molar teeth of *Robustus* continued to enlarge (until they became veritable grinding millstones), the molars of the meat eaters reversed their growth and became smaller. Meat is not only a more concentrated source of energy than vegetables, but an excellent source of protein with a natural balance of all the essential amino acids. Meat is not only more concentrated than vegetables, it takes less time to eat. Meat eaters don't have to spend most their waking hours munching.

Our ancestors obtained most of their meat by scavenging, not hunting. Scavenging on the open savanna in competition with buzzards, jackals, hyenas, and lions was an extraordinarily brave move for these reclusive, tree-hiding, bashful apes. Such bravery could only have resulted from extreme desperation, the gradual disappearance of their trees. It is easy to envisage a group of hominids living in an isolated and dwindling pocket of woodlands, surrounded by vast stretches of open, treeless savanna. Year by year trees got sparser, the pocket smaller. Meat and vegetables replaced fruit. Increasing time was spent on the savanna, further and further from the closest trees. Those most alert to the ways of the open savanna were increasingly favored, strongly selected for generation by generation. In the end, as the last trees disappeared, our ancestors were ready to survive on the treeless savanna. Two hundred million years after taking shelter in the trees they were abandoned.

As the trees disappeared, there were compelling reasons why an even larger brain would be worthwhile. Larger hominid groups had a survival advantage on the open savanna. A hundred, stone-throwing, screaming hominids must have given even lions pause. But keeping larger groups together socially took larger brains. There were more interactions. Politics was more complex. Furthermore, savanna hominids, with their stone tools, became ever more reliant on keeping cultural traditions alive. Understanding and outwitting predators and other scavengers in a clever, coordinated fashion would have benefitted from a large brain. Being neither large nor fierce, we had to rely on our wits and social coordination.

Not only were large brains desperately needed, but with the new energy-rich meat diet, they were affordable. Chimpanzees and hominid frugivores could not afford large brains -- they would have to eat fruit nearly constantly to support them -- but meat eaters could.

There was, however, a roadblock on the path to large brains. The pelvic opening had evolved for a chimpanzee-sized brain. Larger heads simply would not fit, at birth, through the pelvic opening, nor was it in the evolutionary cards to significantly increase the size of the opening. That would have required a major evolutionary restructuring, taking millions of years. As the last trees disappeared, our ancestors didn't have millions of years. They needed a larger brain and needed it fast! (Just metaphorically, for in reality, evolution has no foresight.)

There was a simple solution: have the brain keep on growing after birth. A relatively straight-forward change in a few developmental timing genes could do the trick. But this would also extend the prenatal period of infant helplessness after birth. The newborn babies would not be able to cling to their mothers as they climbed the trees. Ah, but the trees were essentially gone! This was no longer a constraint. There was nothing left to climb. Thus it was that as the last trees disappeared, our ancestor's brain size took a giant leap upward, the newborns became babes in their mother's arms as we took up life on the open savanna. Steven Stanley, whose insightful book, *Children of the Ice Age*, developed this theme, called it the "terrestrial imperative." The terrestrial imperative explains why, after not much happening for a few million years, large brains and stone tools suddenly appeared together in the record shortly after the start of the Ice Age.

Forced to make a living on the open savanna, our ancestors rapidly adopted savanna ways. The large game they scavenged on the savanna had very tough hides. These carcasses were not something you just tear apart with chimpanzee hands and teeth and pass about. To surmount this difficulty, our hominid ancestors drew on another stock (if only occasionally used) chimpanzee behavior. This was the manufacture and use of tools to secure food that was otherwise difficult or impossible to obtain. As discussed earlier, various tribes of jungle chimpanzees, to this day, use specially trimmed twigs to extract termites from their nests, rocks to crush nuts, and so on. Our ancestors, in their moment of need, simply amplified this venerable trait. As before, specific tool making and use traditions were passed from one generation to the next by way of cultural imitation.

It may be hard to believe that simple broken stones can cut through tough hides, even thick elephant hides, but this is indeed the case as has been amply demonstrated by Kathy Schick and Nicholas Toth. In *Making Silent Stones Speak*, they describe their "experimental anthropology." They went to Africa and did themselves what our ancestors had done. Not only did they cut through tough hides with simple broken stones, but they dismembered large carcasses and carried them off in manageable pieces before lions and other carnivores arrived. This was, presumably, the original meaning of the phrase, "cut and run."

Microscopic analysis of the earliest stone tools and the bones associated with them reveal scratch marks on the tools (and cut marks on the bones) consistent with their use for cutting meat . What is even more illuminating is that many of the bones also had carnivore teeth marks on them. These marks were overlaid by the subsequent stone tool scratch marks. Carnivores had made the kill. We were the scavengers. The largest number of bones found with early stone tools were those of antelope, presumably Vrba's new suite of species that appeared on the savanna about the same time we did.

As fruit trees disappeared, our ancestors went beyond meat eating and simple tool use as occasional, supplemental behaviors. These behaviors became the essence of their survival strategy on the open savanna, the usual way of food-getting. This shift from the occasional and supplemental to the usual and regular set us on an altogether new path, one which clearly shows up in the fossil and artifactual records. Our ancestors' teeth and jaws adapted to a meat diet. Our oral hardware did not follow the route of the massive grinding machines of our *robustus* cousins. Our bodies stayed smaller, more mobile. *Homo* was both a long distance walker and a runner. Sparse hair (except on the top of heads) and profuse sweating allowed us to go long distances during the day in the heat of the treeless savanna. Our simple tools, really just broken stones, began to show up in great numbers, often concentrated at what were probably camping or butchering sites some distance from the places such stones naturally occurred. *Homo* had arrived.

Ecologically, our ancestors entered new niches. They joined the meat-eating lions and hyenas as carnivores and scavengers on the African savanna. In making culturally-based tool manufacture and use a necessary part of their lives, our ancestors crossed the divide from a predominantly genetic to what would become, eventually, a predominantly cultural world. The cultural transmission of extra-genetic information between generations became the key to their survival. A new life force on this planet was released. Hominids, ever so slowly at first, began slipping out of genetic control.

This primitive culture became an important part of the hominids evolutionary environment. The ability to make and use tools successfully in carnivore niches itself became a selective force in genetic evolution. Individuals with a greater capacity for cultural transmission and retention of information, as well as those with more dexterous hands, had a slight evolutionary advantage. The era of genetic and cultural co-evolution had arrived. A spiral of increasing intelligence, dexterity, and, eventually, sophisticated technology was the outcome.

Tools and their manufacture and use were not the entire story, however. The simple tools adopted some 2.5 million years ago stayed essentially the same for almost a million years. No noticeable improvements were made; no new types of stone tools emerged. It was just the same old thing for many thousands of generations. These tools, sometimes called the Oldowan industry, were, in reality, just simple broken or flaked stones. Scottish-American anthropologist William McGrew compared tools used to secure food by Tanzanian chimpanzees with those of the recent Tasmanian Aborigines. He concluded that the level of sophistication was similar. As mentioned earlier, Nicholas Toth and Sue Savage-Rumbaugh taught Kanzi, a captive bonobo, to make and use simple stone tools. Thus stone tools, in and of themselves, may not have been nearly as important in the emergence of *Homo* as was our move out into the open savanna, freeing our brain to continue its rapid growth after birth, while challenging us to survive in competition with lions and hyenas. Anthropologists increasingly believe that changes in social organization that adapted our ancestors to their new way of life on the savannas were more important than stone tool use in sustaining the evolutionary spiral that produced the remarkably rapid growth of our brain.

Though stone tools remained little changed for a million years, hominid brains continued their rapid growth, from a normal chimpanzee size of about 450 cubic centimeters, about apple size, to a full grapefruit size of 800 cubic centimeters -- an unprecedented near doubling. This is an extraordinarily rapid pace for genetic evolution and, as mentioned earlier, brains are expensive metabolically.

Survival on the savanna required large, cooperative groups, a division of labor, and extensive sharing. Pregnant females or those carrying babies did not make good scavengers or hunters although they made good gatherers. Most calories still came from gathered tubers, seeds, and the like, not meat. Scavenging and hunting required male cooperation. With competition between males deemphasized, males and females became more equal in size. As brains grew larger and the period of postnatal helplessness increased, females increasingly preferred helpful males who stayed with them. Pair bonding began.

Eventually the advantages of an ever larger brain were offset by its increased metabolic costs and the lengthening period of infant helplessness. At this point, the sudden spurt of brain growth fell off. *Homo habilis* had arrived.

With modest home-front success, *Homo erectus*, a slightly advanced form of *Homo habilis*, spilled out of Africa into the southern regions of Asia, venturing as far east as present-day Beijing. Although the *erectus* that stayed in Africa were soon to develop improved stone tools, those that left, early on, for the east still took the old, original Oldowin tools with them. Amazingly, a secluded pocket of *erectus* that settled in Borneo survived to 50 thousand years ago, perhaps only 25 thousand years ago. *Erectus* lasted longer than any other hominid, almost 2 million years.

Meanwhile, back in Africa, the home-front *erectus* developed advanced stone tools, the Acheulean. Beyond broken stones and sharp chips, the Acheulean tools included carefully shaped hand axes. As before, once the new tools appeared they didn't noticeably change again for a million years. At some point in time, we are not sure when, *Homo erectus*, unique among the animals, learned to use fire. In various manifestations, fire, in the hands of *Homo erectus* and its descendants, was to change the planet, transforming ecosystems and releasing vast quantities of energy. *Homo erectus* may also have been responsible for the development (or at least the initiation) of that other uniquely human capability, complex language.

When did our ancestors develop language? The experts remain quite divided on this question, with some suggesting that language, or at least some sort of proto-language, had evolved by 2.5 million years ago, while other experts insist that it may have been invented as recently as 40 thousand years ago. Many different animals communicate with each other, and social animals are especially communicative. Ants, as discussed earlier, are compulsive chemical communicators; their colonies of millions are built on the cooperation inherent in their common genetic behavioral programs and, critically, a couple of dozen chemical words. Also, as mentioned earlier, chimpanzees and bonobos (such as Kanzi) can be taught the meaning of a hundred or so symbols. They can creatively use two or three of these symbols together to create very simple statements similar to those of a two year old human child. In the entire animal kingdom, however, nothing approaches the infinite generativity of human symbolic language.

Spoken language isn't the only means of communication between individuals. We often use gestures. Gestures are an almost automatic supplement to speaking; it is difficult to talk without gesturing. Communication between individuals also takes place when one demonstrates to another how to do something. For *Homo*, tools and how they are made has always been a special category of communication between individuals and between generations. Since we left the safety of the trees and took up life on the open plains, our lives have depended on tools, on not losing the continuity of this knowledge between generations.

Having suggested what language is not, much remains that is crucial to language. This is especially true when language is combined with the ability to plan ahead, to think about and discuss things "off line" before they actually happen, to coordinate actions, and to analyze and discuss what happened after the fact (the better to do it next time). Language would have been a great boon to a species whose very survival depended on cooperative, coordinated group actions. Thus the evolutionary pressure for development of language, coordination, and planning must have been immense. One suspects that these may have been key reasons why larger brains were advantageous.

Creating a larger but structurally similar brain, as *Homo habilis* apparently did, turns out, genetically, not to be too difficult. A change in just a few developmental timing genes can cause the brain grow for a longer duration than the rest of the body. Changing the structure of the brain is much more difficult, however. Those areas of our brain that are different structurally from chimpanzees are primarily involved with the generation and understanding of speech, and with planning and other complex mental tasks. As these specialized areas bump up against the outside surface of the brain, one would hope that their development over time could be traced in the endocasts made from fossil skulls. In practice this has proven difficult. Conclusions based on such evidence remain, at best, highly controversial.

Besides brain specializations to accommodate speech, our larynx has been evolutionarily relocated and redesigned relative to that of chimpanzees. In the early evolution of land animals, food and air passages crossed. The larynx evolved in terrestrial vertebrates so that animals wouldn't choke on their food while breathing. For them it works well. Animals can actually eat and breathe at the same time. All animals, that is, except us. Our larynx was reconfigured to accommodate speech, to rapidly produce a wide range of sounds. This rather slipshod redesign allows food to enter our windpipe should we breath while eating. Evolutionary selection pressures for improved speech were simply much stronger than the penalties of a few deaths from directing food down the wrong pipe.

As with structural changes in the brain, structural changes in the larynx took quite some time. Although mynah birds and parrots appear to have solved such problems in a more compact, elegant fashion, evolution is always constrained make do with what it has, jury-rigging new adaptations from available components.

Once language got started, even simple language, there was bound to be considerable selection pressure for mating with those who spoke (or gestured) a bit better and were therefore more socially adept. There would have been other selective advantages for improved vocal skills, such as increased political power, better care of children, and more seductive sweet nothings whispered into ears. Many rapid changes in animals are due to out-of-control arms races between two different species (such as the speed of a predator and its primary prey). With our ancestors, the intelligence arms race may have been a family affair. Modern humans, as a result, are fast talkers. We can produce distinct sound segments at the amazing rate of twenty-five per second.

While *Homo erectus* was making its debut and evolving onward, what of our fellow eastern apes? What was the gorilla-like *robustus* up to? *Robustus*, it turns out, was also getting smarter, but not nearly as fast, finally going extinct about 1 million years ago. Presumably *robustus* lost out in the competition with our much smarter, chatty ancestors. Perhaps, as suggested by Stephen Boyden, in the long run there may be ecological room for only one species with the capacity for culture. If this were the case, then the first species to acquire significant culture would eventually eliminate any near competitors, nipping any newly appearing species in the bud. *Robustus* might also have gotten the squeeze from the fast-rising and increasingly successful baboons that had invaded their ground vegetable niche.

About 900 thousand years ago, after a million years of relative stasis with *Homo erectus*, our lineage finally came out with a new physique and tool kit as the second great pulse of the Ice Age descended on the Earth. The new body was close enough to ourselves to get the *sapiens* nod. We call them *Archaic sapiens*, the most famous (but not the most typical) being the Neanderthals. Based on large stone flakes, the latest hominid toolkit contained more than sixty kinds of implements. As before, the latest model was hatched in Africa and, as before, it eventually headed north and east out of Africa, this time directly into the teeth of the deepening Ice Age.

Living in Europe in the second half of the last Ice Age, Neanderthals were, in spite of their reputation, highly advanced. Their specialized culture allowed them to thrive in a cold climate. Neanderthals had a brain as large as ours; perhaps slightly larger, although they still used a toolkit similar to the earlier Acheulean kit with its classic hand axes. Neanderthals most likely had at least gestural language, but their spoken language may have been rudimentary. However, an advanced model they were nevertheless.

Finally, sometime less then 200 thousand years ago, biologically modern *Homo sapiens* made its African debut. The new hominid model was eventually followed, after some delay as usual, by improved tools. Its -- excuse me, our -- new toolkit exploited narrow-blade instead of wide-blade technology. Instead of sixty types of implements, we had well over a hundred. Several African sites show the new, considerably advanced tools by 90 thousand years ago. The physical and mental makeup of these ancestors was probably identical to ours today. Properly dressed and shorn, any *Homo sapiens* from 100 thousand years ago would today raise not an eyebrow in any cosmopolitan city. They were thoroughly modern. Any one of them could have learned to fly a 747. With them, the present human condition was reached. Significant changes since then have all been cultural, not genetic. The recent origin of modern humanity and our genetic similarity can be grasped when one realizes that genetic diversity between humans is only one fiftieth of the genetic diversity between chimpanzees. Human racial differences are almost completely insignificant.

Soon after their debut in Africa, the latest human model headed out from the homeland. Reaching the Middle East, always the crossroads of the world, some *sapiens* pressed onward into Asia, while others headed west into Europe, still in the grip of an Ice Age. By 60 thousand years ago *sapiens* had reached Australia. Starting 50 thousand years ago in eastern Europe, and ending 30 thousand years ago in western Europe, Neanderthals were systematically replaced (exterminated?) by the *sapiens* invader. Twenty thousand years after cohabitation began, the Neanderthals were gone. Results were similar whenever *Archaic sapiens* encountered modern humans; they simply disappeared. The reason is not known. Perhaps Neanderthals and other *Archaic sapiens* lacked a fully modern spoken language, primarily relying on more primitive gestures for communication. If so, our ancestors may have attacked at night when the spoken word could coordinate the attackers but the gestural word could not be used by the defenders. For whatever reason, none survived.

Beginning about 40 thousand years ago in western Europe, tools of the very finest form appeared. Delicately struck in antler as well as stone, these implements were often, quite literally, works of art. Cave painting, sculpturing, and engraving were in evidence although perhaps not yet common. Wherever *sapiens* ventured can be found similar displays of the modern mind at work. The era of long-lasting static toolkits had come to an end after a two million year hegemony. From this point on, continuous change became the norm.

The reason behind this explosion of sophistication is not known. Perhaps it is more apparent than real, being preceded by similar but rarer examples not yet found. More likely, there seems to always have been a delay of a 100 thousand years or so between a new physical hominid species and a new tool tradition. One suspects that this delay may be due to social reorganization taking place. Male-female pair bonding may have reached its modern form. Tribes may have become larger and social customs more complex. Intertribal warfare may have been on the increase. Certainly a delay of 60 - 100 thousand years is not an unreasonable time for cultural evolution to produce new tools and other artifacts. Cultural evolution, while a thousand times faster than genetic evolution, is not instantaneous.

By 50 thousand years ago, *Homo sapiens* had not only made it to Asia, but had crossed over into Australia. By 20 thousand years ago, equipped with finely tailored parkas and other cold-weather gear, *sapiens* invaded arctic Siberia. Shortly thereafter, our species hiked and hunted its way across the Bering Strait land bridge to Alaska. There they were temporarily blocked by a vast ice sheet. About 12 thousand years ago a corridor opened through the glacial ice. In just a thousand years, the ancestors of the indigenous Americans hiked and hunted their way to the very tip of South America. A sea-faring group also left the Asian staging area, settling the South Pacific islands, reaching Hawaii about the time of Socrates, and New Zealand about the time of King Arthur and Sir Lancelot's falling out. In well under a thousand centuries, *Homo sapiens* came to occupy all the land areas on this planet except for Antarctica.

Coincident with or shortly after the explosion of art and new tools some 40 thousand years ago, the spear was invented and, not long after that, the bow and arrow. These weapons allowed our ancestors to kill large game with relative impunity. The normally lethal defenses of large animals could not be brought to bear on hunters equipped with spears. In Africa, Europe, and to some extent Asia, animals had become accustomed to earlier humans hunters, *Archaic sapiens*. The beasts had learned, over many generations, to avoid this dangerous top predator. The large mammals of Africa certainly had an early start in treating primate bipeds with suspicion. They witnessed every increment of our technological advance. But when large animals that had never encountered a hominid were suddenly exposed to spear-equipped humans, the results were predictably and uniformly disastrous. Australia and Americas were, to put it mildly, a turkey shoot. Three-quarters of the large mammalian species were wiped out in each case, driven to total extinction, presumably by human hunting. Our kind -- prior to any civilization -- seems to have caused the greatest extinction of large animals since an asteroid wiped out the dinosaurs some 65 million years ago.

All this was only the beginning. We stood, 10 thousand years ago, on the edge of a sudden shift that would transform human culture and the planet itself with blinding speed.

Further Reading

Allman, William F., 1994, *The Stone Age Present: How Evolution Has Shaped Modern Life.*

Baron-Cohen, Simon, 1995, *Mindblindness: An Essay on Autism and Theory of Mind.*

Bikerton, D., 1995, *Language and Human Behavior.*

Cavalli-Sforza, L. L., 1995, *The Great Human Disporas: A History of Diversity and Evolution.*

Corballis, Michael C., 1991, *The Lopsided Ape: Evolution of the Generative Mind.*

Diamond, Jared, 1992, *The Third Chimpanzee: The Evolution and Future of the Human Animal.*

Fagan, Brian M., 1990, *The Journey from Eden: The Peopling of Our World.*

Falk, Dean, 1992, *Braindance: New Discoveries About Human Origins and Brain Evolution.*

Gibson, Kathleen R., and Tim Ingold, eds., 1993, *Tools, Language, and Cognition In Human Evolution.*

Goldsmith, Timothy H., 1991, *The Biological Roots of Human Nature: Forging Links Between Evolution and Behavior.*

Howells, William, 1993, *Getting Here: The Story of Human Evolution.*

Humphrey, Nicholas, 1992, *A History of Mind.*

Kingdon, Jonathan, 1993, *Self-Made Man: Human Evolution from Eden to Extinction?*

Leakey, Richard, 1992, *Origins Reconsidered: In Search of What Makes Us Human.*

_____, 1994, *The Origin of Humankind.*

Lewin, Roger, 1993, *The Origin of Modern Humans.*

Lieberman, P., 1991, *Uniquely Human: The Evolution of Speech, Thought, and Selfless Behavior.*

Napier, John Russell, 1980, *Hands.*

Pfeiffer, John, 1982, *The Creative Explosion.*

Perner, Joseph, 1991, *Understanding the Representational Mind.*

Pinker, Steven, 1994, *The Language Instinct: How the Mind Creates Language.*

Potts, Rick, 1996, *Human Descent: The Consequences of Ecological Instability.*

Quiatt, Duane, and Junichiro Itani, eds., 1994, *Hominid Culture in Primate Perspective.*

Reynolds, Vernon, and R. E. S. Tanner, 1983, *The Biology of Religion.*

Ridley, Matt, 1993, *The Red Queen: Sex and the Evolution of Human Nature.*

Rue, Loyal, 1994, *By the Grace of Guile: The Role of Deception In Natural History and Human Affairs.*

Sagan, Carl, and Ann Druyan, 1992, *Shadows of Forgotten Ancestors.*

Smith, Eric A., and Bruce Winterhalder, eds., 1992, *Evolutionary Ecology and Human Behavior.*

Schick, Kathy D., and Nicholas Toth, 1994, *Making Silent Stones Speak: Human Evolution and the Dawn of Technology.*

Stanley, Steven M., 1996, *Children of the Ice Age: How a Global Catastrophe Allowed Humans to Evolve.*

Stringer, Christopher, and Clive Gamble, 1993, *In Search of the Neanderthals.*

Tattersall, I., 1995, *The Fossil Trail.*

Tudge, Colin, 1996, *The Time Before History: Five Million years of Human Impact.*

Wilson, Edward O., 1978, *On Human Nature.*

CHAPTER 6

CIVILIZATIONS
The Chimpanzees who became Ants

As planet-wide top predators at the end of the Ice Age, we were an unlikely candidate for breaking through to the superorganism hierarchical level pioneered by the ants and other social insects. Individualists, we lacked the selfless, unquestioning Communist spirit which had brought them superorganism success via the smooth cooperation of thousands or even millions. Furthermore, we *Homo sapiens* only came in two varieties: male and female. We had no inborn caste structure and thus little specialization. Everyone was a generalist. Our only division of labor was based on sex and age. Nor was this all.

We lived in small bands. Gypsies, we moved on as we exhausted local concentrations of food or when our accumulated filth rendered a location unpleasant or unhealthy. Our largest groupings naturally and inevitably split apart if they got too large for everyone to nose into everyone else's business, too large for personal interactions on a day-to-day basis. Massed insect-like togetherness just didn't seem to be our thing.

Worst of all, we were eating at the wrong end of the food chain, the top instead of the bottom. We were rare, as top predators and picky omnivores must always be. Our biomass was minuscule; less than five million of us planet wide. One would have had to look hard to find an animal less suited to becoming a populous, biomass superstar.

Ecologically, we were in the same sort of situation that primitive hunter-gatherer ants were before they discovered herding and farming. Like them, we lacked the internal cellulose-digesting bacteria of cows or termites. The only way we could become numerous was to mimic ants who had tapped the bottom of the food chain via sophisticated control of other species. They let other species digest plants for them (aphids) or grew their own digestible food (fungi). Only if we emulated the herding or farming ants could we grab a more generous portion of the planetary pie. But why would we want to do such a thing?

A popular misconception is that, prior to agriculture and civilization, we had to work hard all day long just to get enough to eat, that we were always on the edge of starvation and were much too busy hunting and gathering to have any free time. At night, so this story continues, our ancestors retreated to the protection of caves. Large fires in their entrances kept the vicious animals at bay, their hungry eyes reflecting our firelight. Then some green-thumbed Einstein discovered what happened if one planted seeds. Almost overnight, agriculture gave us abundant food with little work, giving us the leisure time to form civilizations. As surrounding hunter-gatherers realized the overwhelming benefits of agriculture and civilization, they too quickly joined in. And that, to conclude the story, is why almost the entire planet became civilized in just a few thousand years.

Social anthropologists, who have studied the few remaining hunter-gatherer societies in some detail, paint a starkly different picture. They point out that hunter-gatherers work only two or three hours a day (if their varied and healthful activities can even be called work). In this short time they are able to secure a balanced and nutritious diet high in protein and fiber and low in fat and carbohydrates. They have no incentive to work more than this and rarely do so. This leaves them with what, to us, seems like an inordinate amount of leisure which they use to gossip, tell stories, participate in social activities, and just relax.

Physical anthropologists, for their part, have assessed the health of humanity prior to and after the advent of agriculture and civilization. Based on skeletal remains, they have concluded that our ancestors were well fed prior to agriculture, but poorly fed afterwards.

Finally, rebutting the core of the popular misconception, it seems likely that hunter-gatherers have always been quite aware of the life cycles of the plants and animals about them. They knew what seeds were for.

Thus the popular notion of stone-age humanity is wide of the mark. It might be noted that since the advent of civilization, mobile hunter-gatherers have had many chances to voluntarily take up agriculture and its sedentary ways. Repeatedly they have rejected such opportunities. Hunter-gatherers aren't dummies. They know it takes more work to produce food by herding or farming than it does by hunting and gathering. To make a living, farmers must work harder than hunter-gatherers. Instead of just picking and eating the end result, they must also clear the land, plant seeds, and keep weeds and other would-be consumers (human and non-human) at bay. Worst of all, this hard, repetitious work must be done in open fields in the hot sun. Only by slavery, force of arms, or threat of starvation have indigenous people been brought under the sway of civilized, agricultural life. Australian aborigines, to this day, still insist on their walkabouts.

The hunter-gatherer life style remained well-suited to humanity as long as population density was low and game and other food remained plentiful. Their naturally high protein, low carbohydrate diet, and frequent walking exercise kept their body fat low, resulting in low female fertility (which increases with body fat). Breast feeding each child for four years or more also served to decrease fertility. Nor were mothers who walked a couple of thousand miles every year thrilled with the thought of more than one babe-in-arms at a time; another result of the "terrestrial imperative" we met in the last chapter.

Hunting and gathering was a lifestyle evolved over a couple of million years. It was a lifestyle that was healthy both physically and mentally. Evolution "designed" us to be hunter-gatherers. Why didn't we just stay healthy and happy hunter-gatherers?

The answer, in short, is that we were the unwitting victims of our own success. As mentioned in the previous chapter, beginning about 40 thousand years ago, cultural information began to snowball. A rapid improvement in hunting technology led to an increased reliance on meat from large mammals. Even the very largest animals were no match for our spears and coordinated hunting parties. Our population inevitably rose. But we were killing off large game much faster than it could replace itself. Thus the game we had come to depend on began its steep decline; for many species a decline to extinction.

To make matters worse, we could no longer move on to unoccupied territory. Our snowballing cultural information, by way of boats, cold-weather parkas, and similar cultural contrivances, had already allowed us to populate the entire planet (with the exception of Antarctica).

The archaeological record shows a clear shift from large to small game as the Ice Age drew to a close (slightly later in the New World as it was more recently occupied). Hunters secured less meat per hour of effort with small game than large, but they had little choice. As small game and wild fruits and vegetables became scarce, there was a shift to slash and burn agriculture and to pastoral herding. Finally, as good land for slash and burn agriculture or herding animals became scarce, portions of humanity turned to sedentary agriculture. Given the choice of staying in one place and farming or starvation, we chose to farm. It was as simple as that. Our hunting success had sealed our fate. We had no other option.

Agriculture began about the same time (i.e. within a few thousand years) at widely-spaced locations about the planet. This closeness in timing was probably not due to the diffusion of agricultural knowledge. Such knowledge was already well known. It was due to population pressure. When our combined population began to exceed what the planet could support as hunter-gatherers, we began the necessary shift to agriculture.

There is a two-part, plant-animal strategy to farming. The plant part is to make common those usually rare plants, such as grains, vegetables, and fruits, that we can eat directly. At the same time, we make plants which are not edible to us much less plentiful then they otherwise would be. We physically alter the landscape (clearing, leveling, plowing) to benefit our chosen species. Directly planting the seeds of our favored few, we relentlessly attack any other life which dares encroach on our humanly-created ecosystems.

The animal part of our agricultural strategy is to make common those animals we find useful while, at the same time, making competing animals rare. But why did we bother with animals at all? Wouldn't we have been better off just eating plants at the bottom of the food chain? Why pass scarce food through other animals first, thereby losing most of it? Why eat anything at all off the top of the food chain?

Vegetarians are living proof that we can survive while only eating plants (although fruits and nuts are, in a sense, at the top of the food chain). Survival is one thing. Obtaining a nutritious balance of proteins is another. This is not easy for vegetarians. We had been eating significant amounts of meat for several million years. Because meat is a naturally balanced source of protein, we lost the ability to create several essential amino acids from plant foods. Although we can, via a careful combination of plant foods, obtain these amino acids, meat and animal products provide these essentials in a more direct, convenient, and tasty manner.

Not only was meat our best source of protein, but it came as something of a freebie as cattle, goats, and sheep can exist on plant food that we ourselves cannot use directly. There is the straw and stubble left over from grains, for instance. With most vegetables we only eat certain favored parts. Why let unused portions go to waste? Better to have domesticated animals eat them and provide ourselves with a source of balanced protein by way of their mean, milk, or blood. Similarly there are many parcels of land that are ill-suited for farming but just fine for grazing. Domesticated animals not only produce valuable balanced protein, but are also a source of fiber for clothes, hides for shoes and tents, and fertilizer for gardens. Some of these animals, especially cattle, proved useful for pulling plows and transporting material via sledges.

Together, humans and their domesticated plants and animals form an efficient team. Together we have rapidly taken over the planet, achieving combined biomass stardom by way of our cooperative effort. The biological effectiveness of our agricultural strategy cannot be denied. Nor were we slow in refining our domesticates or experiencing the consequences of agriculture.

Just as we had known about seeds and life cycles from time immemorial, so too had we known about heredity. With the advent of agriculture, we took charge of evolution. The plants we favored for propagation in our ecosystems were those that were easier and faster to grow and cultivate, more convenient to process or store, or had larger edible portions that were tastier, more nutritious, or less toxic. Animal offspring favored were those which were less agile, more passive, had thick furry coats, or could pull larger plows.

Domesticates are not so much wild species held against their will as they are new species that thrive alongside humans in our unique environment; partners in a brave new world order. Their wild relatives have become rare or have been driven to extinction. It may not be noble to be a kept plant or animal, but from an evolutionary viewpoint it gave a few chosen species a shot at planetary stardom.

Sedentary agriculture has been practiced by various species of ants for well over 20 million years. We, however, were the first mammal to engage in such agriculture and the first species of any sort to combine a sizable number of both plant and animal domesticates in the creation of new ecosystems. There were three major consequences of this biologically unique development.

The first was that once an area became agricultural, its population, human and domesticate, boomed. Compared to natural ecosystems harvested by hunter-gatherers, intensive agriculture can easily support a *thousand* times as many people on the same area of land. Being sedentary, mothers no longer need to carry their children about and can bear more children. Furthermore, agriculture generally produces a high carbohydrate, low protein diet which naturally leads to body fat and hence to increased female fertility.

As human and domesticate populations soared, local populations of game and edible wild fruit and tubers plummeted. There is only so much space, water, and sunshine. The path from hunter-gatherer to sedentary agriculturist was one way. Agriculture and its attendant population growth was the ratchet that would propel humanity, like it or not, towards planetary dominance. The logical outcome, unless we become the first species in four billion years to voluntarily restrain its success, that all the natural ecosystems on the planet will be turned into one gigantic human agricultural ecosystem (or we will go bust trying to make it happen).

The second consequence of settled agriculture was a turn for the worse in the lifestyles and health of the great masses of individual humans. Food was no longer an inalienable right, free to all for the picking or hunting. Obtaining food required long hours of hard, monotonous work in the hot sun. The food so obtained was less varied and nutritious. It contained less animal protein and more carbohydrates. Grains are a concentrated energy source that are relatively easy to grow. Grains store well; nature evolved them to be hardy seeds. While grain is calorie rich, it is deficient in a number of essential amino acids and other nutrients. The malnutrition of early farmers is evident in their skeletal remains. Physical anthropologists have established beyond doubt, for instance that rickets became widespread. Humans became noticeably smaller, losing several inches in height (a loss not regained until modern times).

Not only were farmers malnourished, but by living continuously in one place, their accumulated food and filth attracted pests and parasites. Cohabitation with farm animals led to development of many new diseases, and failure to frequently pick up and move to new locations led to continuous reinfections. Again, skeletal remains make it clear that farmers, contrasted to hunter-gatherers, suffered from many diseases.

Sedentary life also had undesirable psychological effects. In cases of serious disputes, people were no longer free to pick up and move elsewhere. There were no easy outs. As population density skyrocketed, the number of individuals one had to interact with quickly rose beyond the modest number we had evolved to comfortably handle. We were frequently forced to deal with strangers.

It is not surprising that farmers remembered their former lifestyles with affection and nostalgia. The Biblical story of our banishment from the plentiful Garden of Eden, forced into earning a living by the sweat of our brow is, anthropologists now inform us, a true story. Unlike ants, which had millions of years to genetically adjust themselves, we became highly successful agriculturists almost overnight. Only one step removed from easy-going, fruit-eating jungle chimpanzees, we remain hunter-gatherers at our very core. It is the lifestyle for which we are evolutionarily fit.

The third and final major consequence of sedentary agriculture was the generation of a modest surplus of food. Hunter-gatherers don't work more than they need to obtain their daily food. They have no reason to. They were mobile and couldn't take food with them (or much of anything else for that matter). Non-mobile farmers, on the other hand, could save up for a rainy day. This gave them an incentive to work for more than just the food they needed for themselves. They could, under the right circumstances, be enticed generate a surplus. Since agriculture is less labor efficient than hunting and gathering, it could easily have turned out that farmers, working flat out, might have barely been able to grow enough food for themselves, leaving no surplus at all. This was indeed close to being the case. Conversely, it is conceivable that while we had to stay put and work harder, the surpluses might have been sizable, ushering in an instant golden age for all. The practicalities of farming (prior to machines) were such, however, that on the average, nine hard working farming families all living near the edge of subsistence could generate just enough surplus to support one non-agricultural family. We traded, in almost an even swap, the work efficiency of the hunter-gatherer for the space efficiency of sedentary agriculture.

The very slight surplus, small though it was, was our species ticket to superorganism status, to the cities of thousands and then millions, to an accumulation of information the likes of which the planet had never seen before. As the Golden Age of hunting and gathering faded from memory, shoulder-to-the-plow humanity worked to generate the small surplus of food it set aside for the future or to exchange for pottery, tools, or irresistible trinkets made by an emerging group of specialized craftsmen who gathered together in centrally located villages. With no need to pack up and move on, accumulating such goods became desirable. The consumer society was born.

Initially, village chiefs provided a redistribution service that benefitted all concerned. Their representatives collected surpluses from farmers who were doing well, dispensing village-manufactured goods in return, and provided food for those who were in need. The chiefs transferred goods from those who had more to those who had less. They also provided for the common good by storing food for lean times or village celebrations, and by providing for the defense of the village and its associated farms.

This initially equitable arrangement between those that had more and those that had less, was an open invitation for exploitation. It is a simple biological fact that surpluses -- any concentrated food -- attract clever exploiters. It was not surprising, in fact it was to be expected, that as surplus food accumulated in centralized villages, some humans would invent a way to permanently capture these surpluses in a less than equitable manner. Biology would expect no less.

Perhaps such control was relatively benign at first, but clever minds were not blind to the golden opportunities that surplus production created. Increased control led to greater work for the masses and larger surpluses for the few. Once this runaway, positive-feedback process began, it continued until the maximum possible control (and hence the greatest possible surpluses) were generated for the benefit of those in control. Furthermore, those in control were always alert to the possibilities for even greater surpluses than currently achievable. (We still are.) As control increased, transfers increasingly ran from those that had less to those that already had more. As American anthropologist George Cowgill put it, "A degree of exploitation considered criminal by one generation was tolerated by the next, and soon hallowed by elite-inspired ideology as built in to the structure of the cosmos."

This then is how easy-going, laid-back, egalitarian, chimpanzee-like humans were transformed into industrious, hard-working, ant-like cogs in civilized superorganisms. Although humans lacked the self-sacrificing Communistic spirit of ants, near total control could, nevertheless, tap the full potential of the human animal to produce surplus food. Humans could be enticed (or forced) to work from sunup to sun down doing repetitious but productive labor while being fed a diet of cheaply produced grain. The trick was to be among the elite in charge and then stay in charge throughout the generations. The first systems that evolved to achieve such all-encompassing control were the human superorganisms we call city-states. Humanity was being captured by ant-like superorganisms.

It is tempting to think that control by the elite was achieved by direct, crude force of arms, but all governments, even the most despotic have, in the main, relied on voluntary obedience. It is much more efficient to have willing, cooperative, and, occasionally, even enthusiastic subjects than it is to have subjects who feel oppressed, coerced, or unfairly treated. So how did early civilizations convince the farming masses that they should work hard all day long and permanently contribute a generous portion of their hard-won surpluses to the city elite?

First off, real services were provided. The most important was protection via standing armies from the armies of other city-states. Nothing ruins one's day like being captured as a slave or killed outright. Another vital service was the production and distribution of highly useful farm implements; pottery, and the like. Public works, especially the development and maintenance of irrigation systems, were clearly beneficial to all. Trade with distant kingdoms expanded the scope of available materials and products. Finally, the maintenance of law and order, was a real (though often abused) service. Unlike the small face-to-face tribes and bands, we now had to depend on total strangers for protection and many of life's essentials. The state supplied vital new conventions; values, rituals, and laws to replace the lost small-group (I've known you all your life) social pressure.

Besides providing actual services, elites (especially their leaders) provided legitimacy by claims of service, including the claim that inequality was not only necessary, but in the public interest. *The Epic of Gilgamesh*, humanity's most ancient written story, contains political exhortations by Gilgamesh, an early Sumerian King. They are amazingly similar to the speeches of current mass politicians: protection of the weak from the strong; provision for the blind and aged. Human psychology hasn't changed since the dawn of civilization. People want to believe, want to hear the words spoken (even if reality falls short).

Prior to civilizations, astute leaders understood the human tendency to surrender power to a leader, the warm comfort of being dependent. As Ernst Becker has suggested, this stems from our infantile surrender to and dependence on our parents. State rulers surrounded themselves with symbols of power and authority. They encouraged hero worship. They did what they could to convince their subjects that they were wise and benevolent yet powerful father figures, and that the state (large and faceless as it may have been) was a close and warm extended family. Kinship terms and practices were used on to stress the "family" relationship.

There is a human yearning to be assured that physical death is not final, that we belong to something larger than ourselves and our immediate surroundings, that somehow it all makes sense and will, in the end, turn out to be fair, just, and glorious. There are those who claim that early state-sponsored religions were just a facade the elite used to intimidate their subjects and legitimize their privileged access to scarce resources. Certainly there is some truth to this. Religion, however, provided a genuine service, and one suspects that its priests and other practitioners were, in the main, true believers, not charlatans. While it may be true, as Karl Marx cynically suggested, that religion was the convenient, state-sponsored "opiate of the masses," it is also true that, with the advent of agriculture and civilization, the former easy-going hunter-gatherers badly needed an opiate. As civilizations increased life's complexities, conflicts, and maladapted hardships, we desperately needed some way of making sense of it all.

Hanging in the background, prompting the subconscious mind, was the threat of coercion. Knowing that the state was the source of one's food, one's very existence, knowing that the state had the power to punish, even kill those who opposed it, and knowing that there was nowhere else to go (especially for those civilizations surrounded by barren deserts or hostile neighbors), tended to make subjects voluntarily follow the rules. When all else failed, there was as a last resort, the direct use of force (which few states have ever hesitated to apply).

The first human expression of the planet's newest superorganisms were the city-states in Sumer. The intensive agriculture of these city-states took full advantage of the nearly level land in Mesopotamia. For over 200 miles, the Tigris and Euphrates rivers fall only about 100 feet, i.e. only a half-foot per mile. This level land was already devoid of trees, being quite desert like, yet the rivers provided copious quantities of water. The catch was that irrigation channels had to be dug and maintained. This required a high degree of planning and central control.

Irrigation produced heavy yields of grain and other food. Although the food yield per hour of direct labor was relatively higher than earlier agriculture, so too was the labor required to keep the irrigation systems in order and the number of administrators required to manage these complex societies. It was discovered, early on, that castrating domestic cattle produced oxen, a tractable animal that could pull large plows across the level fields with ease. Oxen and plows, when combined with fallowing, promised seemingly perpetual use of the same fields. Fallowing is a neat trick. One allows all the non-domesticated plant species (i.e., weeds) to germinate and grow. Then, before they have a chance to go to seed, they are plowed under, not only killing off the competition to the domesticated plants but, in the process, providing them with nutrients. The combination of fully-controlled human labor, irrigation, and plows and cows yielded large surpluses, an efficient way, in a small, controllable area, of producing food beyond what the farmers needed for bare subsistence. The result was real cities, not just villages. Now the specialized, non-food-producing, parasitic elites could cluster behind protective walls. Large, permanent cities began to dot the plains of Mesopotamia like so many ant hills. These new city-states joined the ant colonies, bee hives, and termites nests as the planet's most complex communities.

The immediate challenge of these city-states was survival. As with ants, the stiffest competition, the greatest threat was from other, similar superorganisms. As with ants, the largest human colonies had a decisive competitive edge, for they could support and field the most numerous "caste" of soldiers. This basic, biological fact about superorganisms was not lost on those in control of the earliest city-states. Their primary goal, in a nutshell, was to control the greatest number of peasants. By so doing they could produce the largest surpluses and hence field the largest armies to protect their city-state, not to mention raiding smaller, weaker neighboring states.

The degree of central control rapidly fell off with distance. Little control was possible beyond a day's walk, so early city-states needed to support as many people as possible within a limited area, with as large a percentage as possible being non farming (even though this percentage was always small). What evolved was inherently inefficient from the viewpoint of individual human labor, but it produced and supported the largest possible armies. Armies and superorganism survival was what it was all about. Gone for the masses were freedom from want, equality, leisure time, and the uncrowded, wide-open spaces of nature. These were replaced by regimentation, poverty, grinding work, disease and, for many, slavery. Still chimpanzee individualists at heart, this seems sad to us. For ants, hard work, regimentation, and even slavery were old hat. Ants to Sumerians: "Welcome to the world of superorganisms!"

From the biological (but probably humanly unconscious) viewpoint of those in control it all made perfect sense. Already they were thinking like ants. How hard lower "castes" had to work was of little or no consequence. Nor were those in control necessarily being either evil or selfish. They were simply trying to maintain and protect their superorganisms, to prevent their city-states from being devastated by rival city-states. Only a superior army could prevent them, priests and farmers alike, from being killed or captured. There is a certain irony in this, as the farmers who had learned how to control other species themselves fell under control, although by others of the same species.

While it pays well, being in control was never easy. There were lots of greedy hands and hungry mouths between the peasants at the bottom and those in control at the top. The Sumerian priests -- the first bosses -- invented a clever means of ensuring that the valuable surpluses collected from the peasants were indeed channeled to them, that bounty wasn't being siphoned off along the way by unscrupulous underlings, or held onto by upstart farmers who aspired to rise above their assigned lot of bare subsistence.

Sumerian priests initiated the first accounting system. The job of accountants was to keep accurate track of the sheep, cows, and sacks of grain flowing through the system. Accountants used marble-sized clay tokens to stand for the various heads of livestock or jars of grain. The output of a farm or district could be tallied and represented by an appropriate number of different clay tokens, the smallest tokens being single units, larger tokens for ten or sixty, one type of token for sheep, another for cows, and so on. These tokens were placed in a clay jar, the top closed with clay, and the seal of authority rolled across the clay while it was still wet. Stored on an appropriate shelf at the temple, this permanent record helped keep everyone honest, i.e. it helped keep those on the bottom poor and those setting the rules rich. If a question arose, the appropriate jar could always be broken and the tokens counted.

As we all know, records alone do not honest taxpayers make. No dummies, the priests hired the first auditors to keep an eye on the system. A bright Sumerian that ran the audit team for some early priest had the vital job of spot-checking the system. Very shortly after an assessor had counted some farmer's production, and the assessor and farmer had placed their seals on the clay jar, the auditor would, selecting this jar at random from those arriving at the temple, break it open, count the tokens, and then go out to the farm in question and count what was actually there. Any deficiencies, and the wrath of the Gods was summoned full force on the cheaters.

This was a smoothly operating accounting and auditing system, and the priests did well. It was a bit of a pain, however, to break open jars, count tokens, and then reseal them in new jars, not to mention the expense of making new jars. One day, a bright young, efficiency-minded auditor suggested that the tokens, before they were placed in jars, should hereafter be pressed on the outer front surface of the jars. This way the check on the farmer and assessor could be made by the auditor by "reading" the front of the jar without breaking it open. Please note that the work of making the impressions, of filling out this first tax form, was incumbent on the farmer, not the assessor, and certainly not the priest's auditor.

Within a few generations, the temple librarian, perhaps a grandson of the bright, young auditor, was faced with a growing shortage of shelf space for tax jars. In another stroke of efficiency-inspired genius, he made the further eminently reasonable suggestion that tokens and jars be dispensed with altogether. The curved jar fronts contained the same information as tokens within the jar, and jar fronts alone would take up less space as they could easily be stacked. So it was that the first clay tablets, complete with curved fronts and token impressions, came to be.

It was then but a few short steps to replace tokens with a stylus, square up the jar fronts a bit, and add symbols for non-accounting information; ideas, even spoken sounds. The Sumerians eventually employed more than one thousand different symbols in their cuneiform writing system. This complex system was so difficult to master that the first schools were established, some five thousand years ago, for this very purpose. Some of the earliest preserved writing, in fact, comes from student exercises at the school on #1 Broad Street in Uruk. Most occupations were learned through on-the-job training, usually from one's parents, but training scribes required the formality of schools, professional teachers, and the fear of punishment, involving, as it did, long hours of drill and dull lessons. (The more things change, the more they stay the same.) Boring or not, the combination of writing, schools, and stored copies of written material (libraries) began an accumulation of extra-genetic information that transcended localities, generations, even languages. This culturally born information eventually outstripped, in pure volume, all information that had been accumulated genetically for almost four billion years. Clay tax forms and accountants were the spark that ignited the modern age and changed the planet.

Surpluses generated by irrigation and plowing (not to mention careful accounting and auditing) were accumulated in cities as stored grain, and as gold and various other goodies for the priests and other Sumerian elite. This concentrated wealth was irresistible to would-be looters, requiring a permanent and specialized military caste to protect it (and its elite owners). In the competition between city-states, states that needlessly left surpluses with the farmers or squandered them by way of excessive temple building eventually lost out, as other city-states who taxed their farmers more heavily or squandered less had bigger and better equipped armies and, in the end, took over. Prudent Sumerian rulers therefore always taxed their farmers to the max, right down near bare subsistence. The successful rulers ran a tight ship, spending most of their surpluses on the military. The arms race was off and running: horses, chariots, bigger cities, and soon (in another tradition pioneered by the ants, albeit somewhat weakly) multi-city empires.

Sumerian priests originally ran human superorganisms from their city temples which sat at the pinnacles of artificially generated mountains (often the residue of earlier temples). This topophilia speaks to an instinctive affinity of the human mind for equating large size and elevated position with awesome, god-like power. The priests discovered this psychological trick early on. Too many priests and temples, too much religious behavior bled strength away from the military, however. Excessive acquisitiveness of the priestly caste was thus self-defeating. The inevitable result was military instead of religious control, although priests remained among the elite, for they were indispensable for keeping the peasants in line by way of sophisticated but low-cost psychological manipulation. With the military elite now exercising more forceful and direct control of the masses, religions evolved into less onerous institutions that emphasized the use of persuasion rather than force to entice the masses to work hard and willingly for those in control while suffering their many privations with a cheerful spirit. Religions could indeed work miracles!

Around the time of Gilgamesh (2700 BC), the frequency and intensity of warfare in Sumer increased dramatically. This was not so much a result of fighting over limited resources as the ability of a few densely populated, rich states being able to field large armies to overcome others with smaller armies. The rewards, as in ant warfare, were expansion of territory and the capture of slaves and booty. With Gilgamesh, King of Uruk, the first long-distance campaigning began, and cities of Sumer were increasingly walled. By 2500 BC, stone weapons were out and bronze helmets and swords were in. The composite bow and four-wheeled battle wagons drawn by four horses soon appeared (horses were not yet large enough to ride). A warrior culture, separate from the rest of civilized society, began in those times (and has continued to the present). Exclusively masculine, confrontation and violence were rewarded in sharp contrast to the generally peaceful and non-violent life of civilians.

The combination of long-distance campaigning and written communications (which could be sent long distances under the King's seal) allowed a number of city-states to be organized such that between-state infighting was minimized. Sargon of Akkad was the ruler of the first extensive human empire. In the course of over thirty different wars, he forged an empire, corresponding roughly to the borders of modern Iraq, that lasted from 2340 - 2284 BC. His campaigns outside his empire reached Lebanon and southern Turkey.

While empires resulted in entire peoples becoming inferior classes, the dynamics of empire-building mitigated some of the harsher aspects of civilization. Emperors cultivated an image of being wise and benevolent rulers among potential subject populations, thus luring them into the empire with minimal fighting. "Join us and the empire will protect you from overly zealous local tax collectors," they suggested. The emperor usually portrayed himself as a friend of downtrodden farming masses.

Empires never took off among ants, but with humans it was a different story. With empires, the emergence of the modern hierarchy of a humanly-dominated world order can be discerned. Empires controlled vassal states. Within these states, kings controlled the mass of farmers. The farmers, in turn, managed domesticated plants and animals. Some of these domesticated animals, such as cows, contained trillions of cellulose-digesting bacteria. The bacteria ... Hierarchical biology ad infinitum. There was, however, a serious, basic, biological flaw in this unconscious strategy for planetary domination.

The flaw, quite simply, was that the artificial ecosystems we created to favor our own species of domesticated plants and animals (not to mention ourselves), inadvertently happened to also favor a few other species of life. Plants and animals were, in a way, competing for the attention of humans, looking for that free ride to planetary stardom. The only difference between wheat and weeds, for instance, is that we have no use for weeds. Yet both of them have, in a sense, been domesticated by humans. They both thrive in artificial, humanly-created ecosystems. What was particularly unfortunate was that some of these "inadvertent domesticates" viewed our masses of intended domesticated plants and animals as irresistible meals. Worse yet, some of them viewed the high species concentrations of *Homo sapiens* (and our wastes) as a veritable feast.

Although our own genetic evolution was slow, the genetic evolution of fast-breeding bacteria and viruses was rapid indeed. Pathogens quickly took advantage of this new opportunity, civilization. Although civilization turned out to be a poor deal for toiling human masses at the bottom, it was a fabulous deal for microbes and larger parasites, such as grain-eating insects and rats.

Microparasites feed on us, our domesticated plants and animals, our accumulated surpluses, and our accumulated wastes. The world's first riches of garbage were in Sumer, as were the first monocultures -- solid stands of plants -- not to mention granaries chock full of pure food. Rats, insects, and other pests had a field day. And there in Sumer was also the world's first large single-species stands of humans. Human superorganisms were a prolific breeder of unwanted worms, mosquitoes, fungi, bacteria, and viruses.

In even the mild human concentrations of villages, tuberculosis, leprosy, and cholera soon appeared to darken our new existence. But it took massing together in cities to provide human concentrations that promoted the continuous reinfection required by such vicious killers as smallpox and the bubonic plague. It is worth repeating that macro, mini, and micro parasites all required large, permanently emplaced concentrations of humans. Hunter-gatherers, by staying dispersed and ever wandering, avoided accumulated surpluses, garbage, and excrement, and thus did not attract and propagate parasites (or booty-seeking soldiers, for that matter).

Essentially all citizens of Sumer were infected with smallpox and measles while they were still very young. Individuals who survived infancy were immune for life. Although many died young, they did not carry much societal investment to their graves. The steady deaths of Sumerian young was made up for by having more young. As these deaths were spread out over time there was little effect other than immense human misery. When Sumerian civilization came in contact with populations insufficiently large and dense to continually reinfect themselves, civilized diseases spread like wildfire through the non-immune populations, killing off young and old alike. The effect was devastating. Those left alive fell easy prey to the Sumerian armies (which had usually started the infections to begin with). Sumerians, with their nasty diseases, always had a survival advantage over their less civilized neighbors.

This was Sumer's (temporarily) winning combination: domestication of other species, exploitation of the land, breeding like crazy, taxation of workers to bare subsistence, supporting as large an army as possible, and cultivating the nastiest possible diseases. What did this winning combination achieve? Growth ever outward from Sumer and the other early centers of civilization, as humanity began its second great expansion round the globe. This time instead of just large mammals and birds biting the dust, it was entire ecosystems. In the first expansion of *Homo sapiens* as top predator, it was the naivete of large animals unaccustomed to spears that was their undoing. In this second expansion of *Homo sapiens* as superorganisms, it was the naivete of entire ecosystems that was their undoing. But how can entire ecosystems be naive?

Occasionally in natural forests a tree falls, leaving a temporary opening to the sun, or a river cuts a new course, leaving the old river bed exposed. A special group of plants and animals, the opportunists, quickly move in, do their hurried thing, and as the trees reclaim the land, move on to new disruptions elsewhere in the normally undisrupted forest. Civilization is, ecologically, like a falling tree, a cutting river. It makes rents in the normal state of nature. Rather than causing small, temporary disruptions scattered here and there, the onslaught of civilization is a continual, widespread savaging of natural systems via plowed fields, razed forests, over-grazed pastures, expanding cities, and monumental garbage heaps. We have declared war on natural ecosystems.

Now for the clever part: civilized humans, their slave animals, weeds, pests, and diseases-- the whole lot -- coevolved in these continually disturbed environments. They evolved to prosper on widespread disruption. They love disruption! Like a scattering of spores on agar, the fungus of civilization was poised for hegemony. The wholesale conversion of natural to human ecosystems had begun. Just as ant superorganisms out-competed solitary wasps, so human civilizations pushed hunter-gatherers to the periphery.

This civilized combination, built on ecosystems in despair, came to full flower in the Middle East about 3300 BC and also began, mainly independently, in Egypt (3100 BC), India (2000 BC), and China (2000 BC). In the New World, civilizations completely independent from those in the Old World began in Mesoamerica and Peru before the beginning of the AD era. Civilizations were opportunities opened up by surpluses available when humanity took up sedentary agriculture. These surpluses tended to be largest where soils were well suited for long periods of intensive agriculture. It is no coincidence that most civilizations began in river valleys where the rivers themselves periodically replenished the land with fresh nutrients.

Most early civilizations were what Karl Wittfogel termed "hydraulic." They were based on the construction and maintenance of large-scale irrigation systems which required central control and planning. The control of vital water supplies gave the authorities vast power. Extensive lines of transportation and communication were maintained from governmental centers to the cities and the smallest villages. Political power ran in one direction, from top to bottom. Taxes and tributes flowed in the opposite direction. These hydraulic societies, in an age before powerful machines, were dependent on massed human labor for their irrigational canals, roads, and other mammoth construction projects. With such ant-like armies of workers, it is little wonder that most political observers generally considered these societies despotic.

The lives of farmers and construction laborers were always just a notch above bare subsistence. Malnutrition and disease were rampant. The masses were in the same boat as their oxen, subject to commands of the small corps of elite who kept records and managed society. Over time, corruption among the elite grew, public projects were neglected, and canals filled with silt. Agricultural output declined, no longer providing for the peasant masses. Internal revolt or external conquerors would bring a dynasty to an end, but as canals were restored and corruption reduced, hydraulic societies would struggle on. In spite of their immense human misery, these hydraulic civilizations were stable, lasting for thousands of years, longer than any other. This should give us pause. Ant-like forces within our superorganisms may, without great care, always have the upper hand over our chimpanzee nature.

Despotic power did not characterize all civilizations, however. A clear counter example is Athens during the rule of Pericles. This was government through cooperation. The polis, which replaced the monarchy, equated the interests of the state with those of its citizens. This was in the best of chimpanzee traditions. Alas, democracy did not appear, at least at that time, to be very stable.

The New World civilizations were late starters even though people were in place there 11 thousand years ago, just as they were in the Middle East, well before civilizations started anywhere. Perhaps the New World, starting from scratch, was slower in reaching a hunter-gatherer population dense enough to necessitate sedentary agriculture. The main delay in the rise of civilizations is thought, however, to be vital differences in the fauna and flora available for domestication.

In the New World, domesticable herd animals had become extinct, either through climatic change at the end of the last Ice Age or through hunting. The reason similar animals weren't killed off in the Old World was that they had become wary of humans as we slowly improved our hunting capabilities. There was simply no time for such habituation in the New World. Bison in North America were not domesticable and survived primarily because they were not native. They were, in fact, savvy Old World animals who had recently arrived in the New World along with humanity.

The upshot was that there were no draft animals in the new world, not to mention a shortage of meat. Although dogs and turkeys were raised for meat (and guinea pigs in South America), they ate the same foods as humans. In the Old World, ruminants, which supplied most of the meat, ate grass and other food indigestible to humans -- a significant advantage. Furthermore, American Indians had to domesticate maize instead of wheat. Although an excellent grain in the end (which has now been taken up by much of the Old World), maize was initially much more difficult and time consuming to domesticate than wheat, another factor in late start of New World civilizations.

In general, there was a paucity of easily domesticated plants and animals in the New World. The Americas lie in a north-south direction span, crossing many different climatic zones. This makes it difficult for a species domesticated in one place to be useful in others. The Old World, where the first civilizations began, mainly lies in an east-west direction, across somewhat similar climatic zones. Migration, borrowing, and diffusion of domesticates may have been easier. In spite of their later start, New World civilizations soon flourished. Large Maya temple centers were erected between 800 and 400 BC. Soon the largest cities on the planet were in the New World.

As new civilizations sprang up, older ones faded away or ended in spectacular collapse. By 1000 AD, Mesopotamia, the first civilization, had mainly become a desert. Maya temples were also deserted by 1000 AD. Perhaps weeds, diseases, and other unintended "domesticates" weren't the only flaw in our march to take over Earth. Perhaps we were getting too complex for our own good.

One might think that the growth of complexity would come to a halt once optimality were achieved. Not so. States zoom right on past the optimum to levels of complexity detrimental to their continued existence. As pointed out by Joseph A. Tainter in *The Collapse of Complex Societies*, "organizational solutions tend to be cumulative -- once developed, complex social features were rarely dropped." Over time taxes go up, standing armies get bigger, and the rich get richer. This is a common biological phenomenon: evolution often proceeds along the same pathway, rarely dropping an invention, just tacking on new ones.

Supporting a growing load of "macro-parasites" required that more food be produced. Since good land was already in full production, marginal, less productive land had to be brought into cultivation. As complexity grew, more land, food, and raw materials, had to be processed for decreasing returns. Human superorganisms, like the Red Queen in *Alice In Wonderland*, had to run ever faster just to stay in place. Growing complexity, once beyond the optimum, became a millstone around a society's neck.

Human superorganisms can be expected to eventually collapse, sometimes spectacularly so. Their collapse may be hastened by environmental collapse. As

English historian Clive Ponting remarked, it is "not surprising that the first signs of widespread ecological damage emerged in Mesopotamia -- an area where extensive modification to the natural environment had first been made." The intense irrigation agriculture of the Sumerians salted the soil. Agricultural output fell to a third of what it had been at the peak. Fewer soldiers could be supported. The empire was lost. Sumer faded from memory as encroaching sands covered over the remains of humanity's first superorganisms. The city-states of Sumer were, in fact, entirely forgotten until German archaeologists, 150 years ago, began excavating some unusual mounds in the desert. They found clay tablets by the thousands inscribed with a strange, archaic cuneiform script. Eventually it was decoded to reveal the earliest human civilization.

Nor was the sad end of Sumer atypical. The archaeological record is a chronology of grand failures. Given the self-destructive process of deforestation, salination, and desertification, one cannot help but wonder why all complex civilizations haven't long since collapsed? Though collapses permanently ruined many environments, there were three temporary respites that have allowed civilizations, up to our own age, to prosper nevertheless. They are: (1) expansion into fresh, undamaged territories; (2) the development of new energy sources that allow marginal land to be profitably brought into production; and (3) technical developments that further (1) and (2) as well as aid in other ways, such as transporting food from areas of surplus to areas of scarcity. Although Sumer itself was doomed, its superorganism offspring and other independently initiated civilizations have not only stubbornly persisted for five thousand years, but have positively prospered.

While the outcome of the contest between civilizations and hunter-gatherers was never in doubt, what ensued when Old World civilizations met those in the New World is more interesting. Shortly after human hunter-gatherers arrived in the New World, they were cut off from their Asian homelands by rising oceans as dwindling Ice Age glaciers melted. This continental cutoff occurred after humans arrived, but before civilizations, Old World or New, had begun. Thus Old and New World civilized ecosystems evolved separately for almost ten thousand years. Five hundred years ago, in a colossal, ecological experiment, they were brought into sudden and intimate contact.

Old World civilizations had a couple of thousand years head start on the New World. It is not surprising they were the ones with ships and guns. Also, the Old World had large domesticated animals. The New World did not. Cortez arrived on horseback. Most importantly, the Old World had civilized diseases, most of them developed in cohabitation with large domesticated animals. American Indians came to the New World before Old World civilized diseases had developed and were, before Columbus, remarkably disease free (although the New World did donate syphilis to their Old World conquerors).

The triumph of Old World civilizations, the successful team of humans and their domesticated plants and animals, may have been completed planet wide on August 3, 1938, ending over 100 thousand years of the independent existence of *Homo sapiens* as top predator. On this date, as far as we know, the last remaining group of hunter-gatherers of any significant numbers was contacted by the advanced tentacles of the superorganisms had begun in Sumer five thousand years earlier. The Grand Central Valley of Papua-New Guinea lay unsuspected for a century, hidden by high mountains and surrounded by thick jungle. The first pilot to fly over the island was amazed to see over ten thousand campfires. Europeans, quickly hacking their way through the jungle, made the last first contact.

FURTHER READING

Boyden, Stephen, 1987, *Western Civilization in Biological Perspective.*

Cohen, Mark N., 1977, *The Food Crisis in Prehistory: Overpopulation and the Origins of Agriculture.*

_____., 1989, *Health and the Rise of Civilization.*

Costello, Paul, 1993, *World Historians and Their Goals: Twentieth-Century Answers to Modernism.*

Crosby, Alfred W., Jr., 1972, *The Colombian Exchange, Biological and Cultural Consequences of 1492.*

_____, 1986, *Ecological Imperialism, The Biological Expansion of Europe, 900-1900.*

_____, 1994, *Germs, Seeds, and Animals: Studies in Ecological History*.

Haas, Jonathan, 1982, *The Evolution of the Prehistoric State*.

Harris, Marvin, 1977, *Cannibals and Kings: The Origins of Cultures*.

Kramer, Samuel Noah, 1963, *The Sumerians, Their History, Culture, and Character*.

McNeill, William H., 1982, *The Pursuit of Power: Technology, Armed Force, and Society Since A.D. 1000*.

_____, 1992, *The Global Condition: Conquerors, Catastrophes, and Community*.

Nikiforuk, Andrew, 1991, *The Fourth Horseman: A Short History of Epidemics, Plagues, Famines, and Other Scourges*.

Pointing, Clive, 1991, *A Green History of the World*.

Roberts, J. M., 1976, *The Hutchinson History of the World*.

Schele, Linda, and David Freidel, 1990, *A Forest of Kings: The Untold Story of the Ancient Maya*.

Serpell, James, 1986, *In the Company of Animals: A Study of Human-Animal Relationships*.

White, Lynn, Jr., 1962, *Medieval Technology and Social Change*.

CHAPTER 7

MACHINES
The Geese who laid the Golden Eggs

Charles Darwin's *Origin of Species* was first published in 1859. Four years later, Samuel Butler suggested that machines had also developed in an evolutionary manner, one somewhat similar to biological organisms. Machines indeed are animals of a kind, as both animals and machines convert energy into action. Long-term evolutionary trends towards more efficient use of energy and greater complexities of action can be discerned in both. Butler, considering the speed with which biological and machine evolution were progressing, boldly predicted that machines would eventually constitute a new class of life, one that would soon surpass us, relegating their human creators to second-class status.

Human-made artifacts, like their biological cousins, do have had their own evolutionary pathways. Historians of technology, such as George Basalla, emphasize there are few if any all-new, human-made artifacts, each artifact being very much based on its predecessors. As with biological life, technological refinements are evolutionarily accumulated over time. There is, as a result, an evolutionary tree of descent among various types of artifacts, such as tools and machines, although the "made" only extends back a couple of million years compared to life's billions. Since the advent of *Homo sapiens*, the diversity of artifacts has grown so rapidly that it now rivals the diversity of life itself. There are, for instance, several million U.S. patents, just as there are several million animal species; not so different. And at least one major class of artifacts, machines, consume energy. Like life they have a metabolism. As suggested by Butler, we should consider mechanical things to be another kingdom of life. This would seem especially appropriate for energy-consuming machines.

While there are many similarities between evolutionary processes that produce biological organisms and machines, there are also important differences. Biological life adapts itself to the environment. Machines, rather than adapting to their environment, tend to change it (although ants, as we have seen, also significantly re-engineer their environments, not to mention beavers). Unlike biological evolution, which is not intentional, machine evolution is quite intentional. Human designers compete to make machines more efficient, productive, and capable. Nevertheless, both processes are very wasteful. The vast majority of both genetic mutations and technical novelties are doomed to rejection. The pace of technical evolution is orders of magnitude faster than the pace of biological evolution. This is due not only to the intentionality of technological evolution, but to its increasing utilization and efficient dissemination of extra-genetic information by way of spoken language, writing, and now computers and the Internet.

From a biological perspective, the relationship between fuel-burning machines and humans is explicable. It centers on the concept of energy or, more precisely, the expenditure of energy, work. Most hunter-gatherers were smart enough to stop working once they had gathered enough to eat. They normally gathered enough food to sustain themselves in just a few hours, and this gave them generous time for the many human activities they preferred. The real problem with humans was that even when you placed most of them into a tightly controlled peasant class (even slavery) so you could get a full day's work out of them, the skimable surplus generated was still disappointingly small because agriculture, in terms of food produced per working hour, was less human labor efficient than hunting and gathering. It was less efficient because farmers had to do so many additional things besides gathering the end results. Because of this relative inefficiency of agriculture, it has been necessary for over 90% of the people to be hard-working farmers supporting fewer than 10% non-farmers. Of the non-farmers, only a small portion were ever the real elites that truly benefitted from it all. The elite could not be numerous because agriculture and civilizations based on human work just weren't very efficient. Let's face it, humans, even when you worked them to the bone, still ate a lot and were difficult to keep in line.

Machines were refreshingly different. The more recent and prolific machines ate readily available fossil fuels, were delighted to work twenty-four hours a day, and didn't require a police force to keep them in line. One could go on for hours about the civilized virtues of machines. The most revealing aspect of the human-machine relationship is simply this: machines, unlike humans, naturally produce large surpluses, gargantuan surpluses. The key to winning civilizations is generating large surpluses. The larger the surpluses, the larger the armies and economic wealth. As is the case with ants, those with the largest (and best equipped) armies generally win.

Thus it was biologically expected that the first civilizations to effectively enlist fuel-burning machines would generate enormous surpluses and win the domination game, easily enslaving other civilizations victory by victory. By cleverly preventing other civilizations from independently obtaining their own surplus-producing machines (and by controlling both the machines and the fossil fuel they ate) these winning civilizations could effectively control the planet. Thus the human-machine partnership became a more restrictive club than the earlier human-domesticates partnership.

From the viewpoint of peasants, the elites that ran civilizations, that drove humans to their full potential, were just parasites living off the hard work of the peasants -- right in there with rats, grain-eating insects, diseases and other vermin. Elites were, in historian William McNeill's phrase, "macroparasites." Machines, on the other hand, weren't parasites at all. They ate their own food, fossil fuels.

By tapping an altogether new source of energy, machines initiated a new form of metabolism. Machines were like photosynthetic life, which pioneered an unexploited source of energy, sunlight. Being first, photosynthetic life had a field day, generating an explosion of activity and new forms, polluting the planet with waste oxygen and, in general, running out of control. Nor was upstart photosynthetic life ever brought to heel by traditional non-photosynthetic life. Quite the contrary. Other life had to adjust itself to the new photosynthetic world order.

Machines, which have tapped into a vast, previously untouched bonanza of fossil fuels, are the new photosynthetic life, running out of control, polluting and changing the planet forever (or at least having a decent go at it). The primary difference between photosynthetic life and machine life is that sunlight will last longer than fossil fuel, although as fossil fuels run low, machines may power themselves directly on solar energy (and, presumably, do so much more efficiently than plant life's meager 2%).

The explosive rise to dominance of the machines that feed on fossil fuels is central to understanding modern humanity and to an appreciation of our current predicament. With machines and their gargantuan surpluses, the always exploited masses of humanity began to dream about how they too could live the good life of the elite. How this dream has actually been realized, at least for the civilizations with the machines, is the story of this chapter (as is the story of why it has not been realized for all societies). It is also the story of why, even for the most fortunate, the reality has, increasingly, developed a few nightmarish touches, for highly productive machines have their own peculiar and somewhat alarming downsides.

Until recently, civilizations depended on forced human labor -- human slaves or coolies -- for most hard work. Humans provided the brute force it took to grind grain, raise water to irrigate fields, and cut wood. Then, primarily in Western Europe, starting almost a thousand years ago, a civilization arose that was based on non-human power, primarily water power supplemented by wind and animal power, to accomplish tasks formerly preformed by human slaves. The motive power of running water was captured by sophisticated water wheels that powered an increasing number of cleverly designed machines that ground grain, sawed boards, pumped bellows, and accomplished other labor-intensive tasks. The *Doomsday Book* (1086), complied by the Norman conquerors of England mentions over 5 thousand water mills in southern England, almost one for every fifty households. Windmills, beginning in the twelfth century, sprouted up like spring mushrooms throughout Western Europe.

Why was the West particularly fertile ground for proliferation of these water- and wind-driven machines? Why did the West encourage the evolution of machine life? Why were the crude and often illiterate peasants of Northern

Europe more favorably inclined towards the evolution of machines as compared to the well-educated and highly refined intellectuals of classic Greece and Rome who preceded them? Three reasons, expanded on below, stand out: (1) a pragmatic attitude; (2) decentralized governments; and (3) stiff competition between the various countries of Western Europe.

The pragmatic attitude of the West was -- let's face it -- a low-class attitude. In classic civilizations, the educated elite did not lower themselves with the physical work involved in practical day-to-day living. Such mere practicalities belonged to the lower classes. In early medieval Europe, after the demise of the Roman Empire, the only intellectuals left were monks. Given the lower-class origins of Roman Christianity, it is not surprising that they believed that physical labor directed towards practical ends, far from being degrading, was actually virtuous. As Benedictine might have suggested, weeding the garden also freed the soul of weeds. Nor were the Western Christians reticent about the rewards from nature that practical work could yield. Thomas Aquinas voiced the widespread Christian belief that the very reason for nature was to serve humanity. Therefore the manipulation of nature for economic ends was virtuous. In so doing we were helping God implement His plan. Far from being reticent, the practical West aggressively pursued practical, material progress with a religious, even missionary zeal.

Nor did the low-class West have any sense of shame when it came to foreign ideas. All that mattered was that they worked. Medieval technology was carried forward by practical peasants, stone masons, lumber jacks, and miners with the modest intent of bettering their humble existence. Unlike other civilized traditions, such as Islam, where novelty was scorned, even thought to be evil, the West, particularly during and after the Renaissance, actively sought out novelty. The fanciful design of new variations on existing machines became a respectable intellectual occupation pursued by dreamers who thought up new, often physically unrealizable machines. Leonardo da Vinci (1452-1519) was the epitome of the playful, intellectual creator of such "paper" machines, though most of his designs could not be realized -- at least in his day.

The second reason why the West provided a favorable environment for the evolution of machines was their decentralized governments. Innovators are, almost by definition, unconventional people. Technical evolution does best in those societies where non-conformists are tolerated. Strong centralized governments are dedicated to the status quo, to a conservative conformity which stifles creativity. Looser, decentralized governments are less hostile towards the eccentric. Unlike the centralized hydraulic societies of the Near East and Asia, Europeans, who depended on rain, not irrigation, for water, had less need to centralize. After the demise of the Roman Empire, landed nobles and cities successfully resisted large-scale recentralization. What emerged was pluralistic societies with autonomous spheres of politics, religion, education, arts, commerce, and trade. The rights of kings and centralized governments were curtailed, as by the Magna Carta.

Perhaps of greatest significance for the evolution of machines, Europe remained divided between a number of highly competitive states. Competition was not only military, but economic. National leaders understood that hostility towards innovation eventually translated into economic loss. Thus many European countries encouraged innovation by way of patents, grants, prizes, and medals. While books were occasionally burned and machines smashed by irate mobs, as long as a few European countries remained creative, the others were eventually forced to follow. Modern Japan also grew out of a politically diverse, feudal society.

What emerged in Western Europe was the free enterprise system we call capitalism. Governments no longer set prices or controlled production and distribution. Individuals and firms were free to make these decisions for themselves. Previously, individual prominence was gained through governmental, military, or ecclesiastical careers. The brightest, most capable, and well educated naturally shunned "low class" economic activities. Increasingly, however, merchants that produced and distributed the goods of daily life gained prominence with their accumulations of wealth. Merchants in England and Holland became well represented in parliaments. It is no coincidence that these two countries were early leaders in the accumulation of mercantile wealth.

The individual capitalists who were most handsomely rewarded were those who, in the best of Western traditions, innovatively improved the life styles of the numerous lower classes, not the wealthy few. Economic power resided with large markets. The technical advances which benefitted the masses also penalized the rich whose outmoded production techniques were being displaced, their large investments and livelihoods ruined. In most societies, such vested interests were protected. Innovations were suppressed to maintain the status quo. In the highly competitive West, governments were anxious to spur economic progress, so they protected innovators from the retaliation of their rich and vocal victims, from those they often put out of business. The upshot was that firms which successfully brought innovations to market were handsomely rewarded. Those that failed to innovate, who rested on their laurels, were severely penalized. A more fertile ground for expanding trade and the evolution of machines is hard to imagine. Western civilization was now poised for its planetary bid. But why hadn't China, the most sophisticated, technically advanced civilization on Earth already taken this path?

China, after all, was the first civilization to invent the basic components of modern technology. Its sophisticated use of iron, clever machines, clocks driven by water power, printing of many books from beautifully carved wooden blocks, and the invention of rockets and many other marvels spoke well of its innovation and creativity. The Chinese, unlike the environmentally exploitative and aggressive Westerners, stressed harmony between humanity and nature, as well as harmony within their unified kingdom. They weren't goaded by constant competition from other countries and had no need to squeeze every last possible economic advantage from their unquestionable inventiveness. The story of Admiral Chen Ho is illustrative:

A hundred years before Columbus, the Chinese built a large fleet of huge seagoing junks that dwarfed anything Europeans had ever built. With an army of thousands aboard, Chen Ho voyaged as far as Africa. But rather then going on to establish trade, the Chinese turned inwards again when the Emperor decreed that seagoing junks with more than two masts were forbidden. Soon the massive shipyards were closed.

The West's reaction to the voyages of Columbus (and other early explorers) could not have been more different. In the competitive, exploitative West, the scramble began immediately to build more ships, claim new territory for the mother countries (completely ignoring the sparse, indigenous populations), and extract as much economic wealth from the new lands as possible. The Spanish went directly for the gold, while the ever practical English planted cotton in Virginia and the other new colonies of the American South.

Although the use of water-powered machines to process linen and wool had been increasing in England in the early eighteenth century, it was the processing of cotton into finished fabrics that was to shift the slow evolution of machines into high gear. Cotton, compared to linen or wool, provides better ventilation, is easier to wash, and absorbs dyes and printed patterns well. It is also relatively easy to grow. The difficulty with cotton was that it took almost 50 thousand hours of human work to spin 100 lbs. of cotton by hand. Even then the result was of low strength and poor quality. In 1764, James Hargreaves invented the spinning jenny, a machine that could spin cotton, although not of high quality. Samuel Crompton came out with the mule in 1779, and it spun cotton thread that was finer, stronger, and more uniform than the best linen or wool. It took the mule just 300 hours to spin 100 lbs. of cotton. Finally, in 1785, Edmund Cartwright's power loom transformed cotton thread into fabric, completing the mechanization of cotton textile production.

Cotton exploded. The forested lands of the American southern colonies were transformed almost overnight into vast plantations which grew cotton originally native to Egypt, Asia, and Mexico. Millions of slaves were imported from Africa to work these plantations. A constant stream of ships moved millions of bales of cotton to machines in factories that sprang up along the fast-running streams and rivers of northern England. Supplier of low cost but high quality cotton textiles to the world, England became increasingly wealthy. The cheapest labor in the world couldn't compete with English cotton machines. The industrial revolution was off and running.

Besides the West's innovative spirit and capitalistic zeal, both discussed above, there were three additional keys to the industrial revolution: (1) machines to make machines; (2) a ready supply of low cost, compliant labor; and (3) a new source of motive power beyond falling water.

First, the industrial revolution simply would not have been possible without machines that made machines; lathes, milling machines, and screw-cutting machines that transformed designs into precise, repeatable, metal reality. Machine life has its own peculiar means of reproduction. The essence of the industrial revolution was the substitution of high-productivity machines for low-productivity humans. This required that machines be dependably grown from iron and steel in sizable quantities.

Unlike the leisurely pace of cottage-industry production prior to machines, machines produced textiles and other goods at a prodigious rate. It took a virtual army of hard working, on-the-spot people to tend them and keep them fully supplied. The age of machines had begun. Noisy, dusty factories demanded long hours of mind-numbing, repetitive labor by workers who quickly numbered in the millions. They were offered incredibly low wages. There were always plenty of takers, because a growing number of people were no longer needed in the increasingly efficient English agricultural system. That they were so eager to work in the factories speaks volumes for the lot of the unemployed. Similar to the beginnings of agriculture, it was hard work or starvation. Again, the elites drove humanity towards maximal work. It was a hundred years before the rapidly growing factories brought full employment and hence sufficient worker scarcity to boost wages to a level that allowed factory hands to live decently.

Finally, there was a new motive power for the machines. Water power had its limitations. It was, after all, at the mercy of droughts, floods, and ice. In any event, water power could only be tapped along sizable, fast-moving streams, and these were rapidly filled up with factories.

It was England's good fortune to sit atop a virtual mountain of coal. Coal had occasionally been used in England and elsewhere for heat, light, and even cooking, but was considered inferior to wood, owing to its obnoxious fumes and grimy soot. Coal was little used in England until the forests were nearly gone. Then coal use soared, especially for heating. Easily accessible coal near the surface was soon used up, and had to be extracted from ever deepening mines. A serious problem with such mines was the need to pump out the water that seeped into deep shafts.

This brings us to Thomas Newcomen and the first useful machines that produced motive power from the burning of fossil fuel. The power produced by Newcomen's coal-fed "atmospheric" engines pumped water from mines, beginning in 1712. That date marks the beginning of the fossil fuel machine age. Large, cumbersome, and inefficient, Newcomen's engines were quickly replaced by more compact, versatile steam engines devised by that quintessential mechanical genius, James Watt. Soon steam engines replaced water power as the prime motive force for the rapidly growing manufacturing industry in England.

At the beginning of the nineteenth century, when steam engines were just beginning to see widespread use, coal consumption was just a fraction of what it was to become. A major step in the development of the steam engine was taken by Richard Trevithick. He began operating steam engines at a pressure of ten atmospheres, a pressure Watt (who used less than two atmospheres) considered dangerously high. Nevertheless, Trevithick's compact, high-pressure steam engines turned out to be safe and also more energy efficient than Watt's machines, getting more work out of a ton of coal. In 1804, a Trevithick steam engine powered a locomotive that hauled itself, ten tons of iron bar, and 70 persons along a nine-mile tramway, thus winning a 500-guinea wager.

From this modest beginning, steam-powered railroads, during the course of the nineteenth century, opened up the interiors of vast continents to the English and other European and American builders of trains, those most beloved of all fossil fuel machines. European and American steamships plied the oceans of the planet, while their expanding factories produced material goods in great abundance. Lighted by low-cost coal gas, factory machines kept a growing industrial work force in dingy factories busy long after natural darkness. By the end of the nineteenth century, steam-power reigned supreme. Coal consumption had soared to 95% of the now greatly expanded human energy consumption, completely dwarfing all other sources of power including human and animal.

The earlier shift from hunting and gathering to agriculture was virtually worldwide, as was the shift to civilization, although those who shifted first tended to dominate late shifters. Widespread manufacture and use of coal-fired machines during the nineteenth century was restricted at first to England and a few other western European countries. Although they were soon joined by the United States and Japan (and very much later by a few others), most of the world remained pre-industrial, which was advantageous to the newly emerged industrial countries. By controlling manufacture of surplus-producing, fossil-fuel-burning machines, the West and Japan could control the planet. Kept in a pre-industrial, agricultural state (often with the able help of natives trained at the universities of the colonial powers), colonies could be forced to be the suppliers of food and raw materials, transported by machine-powered trains and ships to the industrial countries where the food was consumed and the raw materials processed in the factories powered by other fossil-fuel machines. A small portion of the finished product was then shipped back to "pay" for the food and raw materials.

Civilizations have always had their ruling elites; now the planet had its ruling civilizations. A few countries controlled all the other countries, working them hard, extracting their surpluses and, as is the way of biologically-efficient parasites, giving little in return. Such control was gained, in some cases, by settlement and by out-reproducing sparsely populated natives (whose numbers were greatly reduced by civilized diseases, not to mention guns). In other cases, control was gained by military force or the raw economic power born of machine-produced surpluses. Recently, control has taken more subtle forms, such as managing international institutions, favorable trade agreements, and the political dispersement of surplus foods.

At the height of the Victorian Age, nineteenth-century England was the epitome of industrialized civilization. England extracted resources from her widespread colonies and dependencies. Her merchants and industrial barons were supported in these endeavors by a naval steamship force with coaling stations throughout the world. Food was even shipped to England from Australia and New Zealand. A few manufactured goods, not to mention large numbers of surplus, unwanted people were sent back in return. With the advent of refrigerated ships in 1876, fresh meat, butter, and tropical fruit from abroad could be had in England and other European nations. The population of Europeans grew rapidly both at home and in their colonies. Their mastery of machinery multiplied their impact far beyond their rapidly growing numbers. As a result, they redrew the planet's map to their own advantage.

The usefulness of steam-powered machines was limited by the difficulty of transporting their bulky fuel long distances from mines, as well as by their large minimum size and slow startup. Their usefulness soared again, however, with the dawn of the electrical age, which was responsible for a vast increase during the twentieth century in the power consumed by humans or, more accurately, by their machines. Electricity was a convenient source of power for many applications, both large and small, though its use was restricted to fixed locations that could be connected, by way of power lines, to central stations where giant coal-fired engines turned massive electric generators.

Coal-eating steam machines turned out to be just the first wave of fossil-fuel-eating machines to sweep the planet. While coal-fired steam engines were responsible for the great success of railroads, such engines were not practical for drawing plows through fields or for hauling people or small loads short distances. Thus, somewhat paradoxically, as steam railroad transport increased, so did the use of horse-drawn plows, combines, carriages, and wagons. Horse ownership peaked in England and the United States at the beginning of the twentieth century. In 1900, an amazing one-third of the farmland in the United States was devoted to growing food for horses. Removal of their residue from the streets of cities and towns had become a major occupation. The stage was thus set for the replacement of both horse-powered agricultural machines and horse-drawn short-haul transport by machines powered with petroleum-burning internal combustion engines. Unlike coal, which was the residue of ancient forests, oil was produced from ancient unicellular sea-life. Solid coal had to be mined. Liquid oil could be conveniently pumped.

While we may be most familiar with the effects internal combustion engines have had on short-distance human travel, from a broader perspective their impact on agriculture has been even more significant. With the advent of internal combustion engines, chain saws and bulldozers cleared and leveled vast areas of land for agricultural use. Gasoline- and diesel-powered engines pumped water to irrigate formerly non-arable land. Petroleum-powered machines were used by farmers in all phases of the agricultural cycle. Tractors pulled the plows that prepared fields. Specialized machines planted seeds. Other machines tended growing plants and harvested mature crops. Oil-burning machines powered vehicles that transported the harvest to rail heads or even to distant final markets. Beginning in 1812, machines even placed food into the sealed, tin-plated cans invented by Peter Druand. Similar vehicles transported fertilizers and other raw materials back to the farm, including refined fossil petroleum products for the many machines. As petroleum-fired machines took over the main work of agriculture, humans were freed from agricultural labor in great numbers and could thus tend the ever-growing number of machines in factories.

In the last hundred years, industrial production has increased by an astounding factor of fifty, most of it since 1950. This was primarily due to the hard work performed by machines. It was also due, in part, to the adaptation of mass industrial production techniques which began with interchangeable rifle parts in the American Civil War and came to full fruition in the manufacture of the black, mobile, petroleum-eating machine we know as the Model T. Henry Ford figured out how to use machines not only to overcome the difficulty most humans have in doing hard work, he also figured out how to simplify the human jobs that remained. This was the birth of the industrial assembly line. In mass-production industry, as in mass-production agriculture, output soared as the number of human workers declined, replaced by the ever more efficient, more automatic machines. The vast number of people released from agriculture, and then from industry, were now employed (if employed at all) in the "third sector," providing a variety of human services to what were now well fed, increasingly affluent industrial societies.

Once machines had taken over the main work in agriculture, the amount of land that could be worked by a single farmer increased by leaps and bounds. The number of farms plummeted. Agricultural output soared as the number of people and animals (horses and mules) working in agriculture dropped. The long unrealized dream of agriculture, of civilization itself, had finally, after more than five thousand years, been realized. At long last, an agricultural civilization could produce more food per hour of human work than the hunter-gatherers who had preceded them. In the end, the mass of civilized humanity could be better off than their hunter-gatherer ancestors. The masses could, at least in theory, become as well or even better off than the elites of yore. All this because fossil-fuel burning machines had taken over the hard work, the real labor -- at least in the industrialized countries and their prime agricultural colonies. Formerly, more than 90% of the humans in civilizations were farmers supporting the fewer than 10% non-farmers. Now, in most industrialized countries, fewer than 10% were farmers supporting the more than 90% who were not. In Australia, as an extreme example, one farmer (and his sizable collection of machine partners) now supports 125 people, 83 of them overseas. This was only possible because we tapped into a new source of energy, oil. We are, to a real extent, increasingly eating oil.

Petroleum-burning machines encouraged large farms with few humans (but lots of machines). They also encouraged a mass-production approach to farming. Instead of a dab of this grown here, and a small patch of that raised over there, entire farms, even entire farming areas, were devoted to the efficient growing of single crops. Large stands of single species of plants were a tempting target for insects and other parasites, however. Hence the need for pesticides and other chemicals (most of which could be continently derived from petroleum).

Varieties of wheat, rice, and other major crops were developed that produced large yields within the machine-dominated, mass-production agricultural environment. These so-called green revolution plants doubled some yields, not because they were more efficient in converting sunlight into plant tissue, but because they increased the human-edible portions of the plants while reducing the non-edible portions. Because the non-edible but life-vital parts of these green revolution plants were attenuated, the plants needed a more extensive, artificial life support system. Compared to previous plants, the green revolution plants needed to be watered more frequently (usually by way of irrigation), needed more easily obtained nutrients (enter chemical fertilizers), and were more susceptible to insect and fungi attack (requiring pesticides and other chemical deterrents). But not to worry, oil could be turned into the necessary fertilizers and pesticides. Oil could be used to pump the irrigation water and transport the chemicals to the farms.

Within a couple of generations, the entire planet, or at least the arable portion of it, was effectively turned into a giant mass producer of plant food for humans, the animals they milked or ate, and the organic fiber and other products they desired. Few rivers actually made it to the sea any longer, their water was used up in irrigation schemes. A human, planetary plumbing system. This vast, nearly planet-wide system of mass production depended on machines and on their fossil fuel feed stocks.

There was little good land left for archaic subsistence farmers, let alone for hunter-gatherers. All their land, even in the poorest countries, was needed for the efficient mass production of the food and fiber desired by the industrialized countries (not to mention their own, rapidly growing populations). The poorest, most populated countries were most desperate for foreign capital, most anxious to grow large cash crops for the controlling countries, and hence most eager to take land away from their own small-time farmers. Ironically, these same countries had to use their cash income to import food from the likes of Australia, United States, and Canada. According to Clive Ponting, fifty countries that had been self sufficient in the 1930s became net importers of food by the 1980s. They used their scarce foreign capital to purchase the fossil-fuel machines that industrial countries made, as well as the fossil fuels they controlled (not to mention the fossil-fuel-based fertilizers and pesticides needed to grow the green revolution cash crops).

The combination of internal combustion engines, petroleum fuels, and the green revolution had increased the control of the industrial countries. While many countries had gained nominal political freedom, it meant little as they fell under tightening economic control. The domination of the West and Japan, as the second wave of machines swept the planet, was fully in place.

Not only were people of the industrial nations more affluent than their ancestors could have imagined, they were also healthier. In the early days of the industrial revolution in England, towns near coal fields and along coal transport routes became increasingly industrialized and crowded with machine-tending workers. There was little sanitation in these factory towns; no sewers or garbage removal. Diseases such as typhus became rampant. The English Parliament appointed a commission to inquire into this state of affairs. Their report, written by Edwin Chadwick in 1842, resulted in the Public Health Act of 1848 and, in 1875, the establishment of a Department of Health.

The new standards set by these actions (and soon emulated by other countries) began to be felt throughout the industrialized world during the early decades of the twentieth century and, somewhat later, throughout most the rest of the world (sanitary standards were easily exported at low cost). The primary initiatives were: (1) provision of pure or treated drinking water; (2) installation of effective sewers; (3) drainage of marshy areas; (4) regular removal of garbage; (5) adequate ventilation of buildings; and (6) government enforcement of all the above.

That the majority of the people in today's industrial societies are well fed, healthy, and have more than adequate housing, clothing, and other basic necessities is, to say the least, a stunning achievement for humanity. This is why, in the West and Japan, we are so grateful for and respectful of civilization (and why much of the rest of the world is so envious). Ever since civilization began, the bulk of civilized humanity had been consigned to lives of grinding poverty just this side of starvation. The sudden turn of good fortune in the industrialized civilizations, which began some two hundred years ago, was a fortuitous confluence of five factors: (1) large and highly productive areas of new land in the Americas, Australia, and New Zealand; (2) highly productive machines burning fossil fuels took over most of the hard work; (3) the mass production techniques in both agriculture and industry that machines and fossil fuels made possible were institutionalized and refined; (4) industrialized countries exercised tight control over non-industrialized countries, keeping them in a pre-industrialized state as food and raw materials were extracted in copious quantities; and (5) of greatest importance, all of this happened suddenly, boosting the supply of food and manufactured goods much faster than the populations of industrial nations could grow.

Every silver lining has its cloud. To begin with, the sanitation measures instituted by industrial societies (and then transferred to much of the rest of the world) did indeed improve human health and reduce early death. The simple sanitation measures outlined above were, in fact, spectacularly successful. The world's population has grown by a factor of four in the century since these public health measures were first instigated. As numbers boomed, the increased amounts of food and manufactured goods that were initially channeled into ever-increasing standards of living now had to be redirected into the basics of life for the rapidly expanding populations.

The vastly increased numbers of humans, some 5.8 billion in 1996, required more food, which required ever more agricultural land. For several hundred years after the Age of Discovery there were almost unlimited new lands, much of it excellent farmland. By the end of the twentieth century, however, all good land had long since been put to use, and ever more marginal land had to be brought into production. Also, the good land already in production had to be farmed with ever greater intensity. Meanwhile, all the agricultural machines, fertilizers, pesticides, food processing equipments, and transportation vehicles consumed vast quantities of energy. Petroleum companies drilled deeper and advanced into the arctic.

The usurpation of marginal lands for farming and the intense use of prime lands accelerated erosion, salination, and desertification (also brought about by deforestation as the last great forests of the planet were cleared away). The total agricultural land, planet wide, has now, for the first time since human agriculture began 10 thousand years ago, begun to decline. Not only is there less agricultural land, but the yields from this land are declining as topsoil continues to be lost and salt and other undesirable residues build up. Insects and other pests have become resistant to the pesticides that used to effectively dispatch them. After 10 thousand years of expanding food supplies, food production appears to have peaked and may even have begun to decline (although human population is still on its way up).

Nor are our machine-induced difficulties limited to agricultural constraints. Hunter-gatherers, with their mobile way of life, had limited possessions, as did the masses of poor civilized peasants. Only a tiny minority of humanity, the elite, could afford a significant accumulation of material wealth. Modern industrial countries are essentially entire populations of civilized elites. Their material possessions in total are immense. Nor have the industrialized masses been content with the basics of life. When everyone has the basics, their status value is lost. Status-value consumption, spurred on by psychologically clever advertisement, has brought on massive over consumption.

Just as mass-production agriculture has hit diminishing returns, so has mass-production industry. With the richest, most conveniently located ores gone, less concentrated minerals must be mined in greater quantities and transported longer distances, the entire process using increasing amounts of energy. Undesirable residues pile up and become more difficult to handle within what is, after all, a closed planetary ecosystem. We are all aware of acid rain, the rise in global temperature, the ozone hole, and the accumulation of toxic and slowly degrading exotic chemicals -- all down-sides of machines and the industrial revolution.

As we near the millennium's end, machines and their capitalist masters reign supreme. Communism, the only real challenge to capitalism, has fallen on hard times. Real communism works for the altruistic ants, but not for somewhat selfish chimpanzees, ourselves included. The West has been joined by a growing number of Western emulators, several who are beating the West at its own game -- the production of vast quantities of goods by machines. The latest wave of machines -- robots -- are concentrated in a single country, Japan.

As global transportation and communication improved, and as Western, capitalistic ways permeated the entire planet, economic competition that began between European states has increasingly become global. The power of governments to control the economy, never very strong in the West, is now totally overshadowed by both the size and logic of efficiency inherent in a global economy.

As we in the West accomplish our chores at the touch of a button, as we luxuriate in a sea of machine-produced goods, as rapidly proliferating chain saws and bulldozers consume the last forests, and as a vast army of machines fill the air, water, and land with their effluents, it has finally dawned on us, the chimpanzees who would be ants, that we might have been had. Machines, initially few, gained our confidence as faithful servants. We gladly helped them evolve and proliferate. But the tables have been turned. They now enslaving us (and consuming the planet to boot). Ralph Waldo Emerson summed up this worry nicely in 1847 (quoted in Robert Heilbroner's *Visions of the Future*):

Things are in the saddle,
And ride mankind.
There are two laws discrete,
Not reconciled, --
Law for man, and law for thing;
The last builds town and fleet,
But it runs wild,
And doth man unking.

FURTHER READING

Basalla, George, 1988, *The Evolution of Technology*.

Jones, E. L., 1987, *The European Miracle: Environments, Economies and Geopolitics In the History of Europe and Asia* (2nd ed).

Jones, Eric, Lionel Frost, and Colin White, 1993, *Coming Full Circle: An Economic History of the Pacific Rim*.

Keegan, John, 1993, *A History of Warfare*.

Landes, David S., 1969, *The Unbound Prometheus: Technological Change and Industrial Development in Western Europe from 1750 to the Present*.

Mazlish, Bruce, 1993, *The Fourth Discontinuity: The Co-evolution of Humans and Machines*.

Mokyr, Joel, 1990, *The Lever of Riches: Technological Creativity and Economic Progress*.

Olmert, Michael, 1992, *The Smithsonian Book of Books*.

Pacey, Arnold, 1983, *The Culture of Technology*.

Rosenberg, Nathan, and L. E. Birdzell, Jr., 1986, *How the West Grew Rich: The Economic Transformation of the Industrial World*.

CHAPTER 8

SCIENTISTS

The Curious Cats who pried open Pandora's Box

Who are we? How did we come to be? These are the insistent questions of many young children, a few inebriated students philosophizing late at night, religious and philosophic scholars, and those curious academics we call scientists. Scientists have allowed their childhood curiosities to get the better of them, pursuing some esoteric question that happened to turn them on over an entire lifetime as if it were the Holy Grail. They are, gratefully, a tiny minority. If we were all scientists, the crops would rot, our machines would fall into disrepair, and civilization itself would quickly grind to a halt for lack of practical attention.

For almost four hundred years, scientists have been on a collective quest to understand who we are and how we came to be. We have always been curious yet this did not, until recently, lead to the cooperative, wholesale prying into nature we call modern science. What changed? Why did modern science arise in the crude West instead of the more intellectually sophisticated civilizations of China or Islam? For that matter, why should any animal, ourselves included, ask who we are or how we came to be, let alone answer such questions?

The chimpanzees, our sister species, are keenly aware of the personalities and current emotions and motives of the other chimpanzees in their group. They keep a careful account of the complex and ever-shifting interrelationships among those they live with day-to-day. Experts at social chess; they have a highly evolved Machiavellian intelligence. They are equipped with mental models of themselves in relation to the other chimpanzees about them. Chimpanzees not only are self-conscious, but conscious of other minds as well and, to a limited extent, can project ahead. "If I do this, he will think that I want ... and will do that, and then she, seeing us together, will think we are ..." These are mental models of social cause and effect.

We have taken this mental modeling somewhat further than our sister chimpanzees, although humans who work with chimpanzees admit to being frequently suckered by chimpanzee tricks. Not only do humans have a larger brain with a more extensive memory, we also have a much more complex and expressive language which, besides improving communication (presumably its main function) may also assist us in our mental modeling of the social realities about us.

Homo sapiens began, certainly not later than 30 thousand years ago and probably much earlier, to apply mental models of human interactions to understanding and explaining non-human animals, even non-living processes. Paintings, carvings, and burials clearly tell us so. Burying the dead in carefully arranged positions with flowers suggests we were cognizant of our own upcoming death and had conceived of an afterlife as a means of coming to terms with the terrifying prospect of our personal demise. Recent findings suggest that Neanderthals, some 100 thousand years ago, buried their dead. Interestingly, elephants are fascinated by their dead, visiting again and again the bones of close relatives, even laying branches over them. Be that as it may, our mental models, having been developed for understanding and predicting human behavior were not entirely applicable to non-human animals, let alone non-living processes. One does the best one can with what one has. To make sense of weather and other physical forces, we charmingly infused them with human attributes, endowing most of nature with human emotions, motives, and purpose.

Infusing non-living things with human purposefulness is normal in human development. The famous French developmental psychologist, Jean Piaget, called this animism, and it is present between ages two and four. It is an over-generalization of a child's early realization that some "objects" (other humans) have mental states. What the developing child eventually learns, is to distinguish between objects with and without mentalities. This distinction is based, to some extent, on whether objects are self propelled or not. Hunter-gatherers believe that many non-moving trees, rocks, etc., have human-like spirits. In this, and in our early animistic religions, it appears that we mapped reality onto our social-chess mental model. This was the natural, inevitable thing to do.

With early civilizations, religion became organized and central to human life. Religion not only rationalized death and the often strange workings of the natural world; it explained why the masses should work so hard at the behest of a few. Early religious systems accomplished this brilliantly. We had no idea of our lengthy, prehistoric, evolutionary past, so early priest-kings suggested, logically, that we were the remnants of a previous Golden Age, that we (especially the priest-kings) were the descendants of Gods. Most myths look back with nostalgia to a golden past, to a lost Garden of Eden. The universe was capricious, and only the constant intercession of Godly priests asking special favors of the Gods could forestall further degradation.

For the next two thousand years, written religions grew in eloquence and detail, linking nature and gods to increasingly sophisticated and comprehensive human-centered explanations. Understandably, priests made no confessions of ignorance or lack of understanding. No critical tradition was discernible.

A few elite Greeks were eventually given license to pursue, within tasteful bounds, lines of inquiry that departed from the traditionally religious, and to openly critique each other's ideas. By substituting reason for superstition, these Greek skeptics and seekers were able to make considerable advances over the abilities of myth and religion to explain the natural world. The Greeks, for example, transformed a collection of various rules of calculation into orderly systems of thought, the most famous being Euclid's *Elements* -- useful to this day. With geometric reasoning, they deduced that the Earth was round. Based on relatively simple measurements, they estimated its diameter with considerable accuracy. Mathematics and science flowered with the Greeks. Although the Greeks relied almost entirely on logical deduction, having little use for experimentation, they demonstrated that human brains, although evolved as aids to survival and propagation in a social environment, could be successfully used, in an appropriately open and critical environment, to ferret out not-at-all-obvious realities about the cosmos.

The Christians who followed the Greeks in the West were not always kindly disposed toward these strange, non-human-centered viewpoints, and purposely destroyed the Greek classics. Thanks to individual Islamic scholars, however, this unusual Greek insight into non-human reality, was preserved nonetheless. Islamic religion required that the Greek views be treated as and foreign, thus defending themselves against the invasion of an alien philosophy with incompatible views. Nevertheless, Islamic scholars not only were allowed to preserve the Greek classics of philosophy and science but, succumbing to occasional temptation, also made many brilliant additions of their own, such as the use of Arabic numerals and the systemization of algebra. Understandably the Islamic world did not pursue this alien line of thought in any vigorous manner; teaching it was banned from institutions of higher learning.

The Chinese, with the largest and most successful civilization in the world, had an extensive educational system for training their bureaucracy. Books were printed in large quantities in China many hundreds of years before printing began in the West. The Chinese educational emphasis was on literary classics that stressed the worthy social ethic of living together in harmony and happiness. Entry to the higher levels of government required extensive knowledge of this literature, especially the Confucian classics. Ancient wisdom was revered. Although brimming with technological genius, China, even more strongly than Islam, kept to a carefully balanced, albeit somewhat human-centered view. Unlike the West, they considered humanity as but one part of the larger whole of life on Earth. Humanity and nature, the Chinese reasoned, should peacefully coexist. Life was a never-ending cycle, and the key was to maintain harmony and balance at all times. The ever tolerant Chinese allowed Confucianism, Buddhism, and Taoism to peacefully coexist side by side.

The Christian West's view could not have been more different. We were separate, above nature, for only we had souls. We had no obligation to the soulless lower animals. Nature was there to serve us. Far from being filled with independent spirits, and hence capriciously unpredictable, there was an underlying single-God-given order and rational to nature. These views led to a distant, objective, and mechanical view of nature. While the Christian-perceived separateness from nature helped science get started, it later became a stumbling block when we tried to understand how we ourselves fitted in with other life. As nature was not filled with spirits, tinkering with it was neither sacrilegious or dangerous. Furthermore, Christianity viewed the Earth as a stage for a divine play which had a definite beginning, a one-way history, and an eventual end. Not cyclic.

Having lost the Greek philosophic and scientific classics, Christianized Western Europe had no need to erect barriers against these powerful alien thoughts. By the time they were translated from the Arabic in the twelfth and thirteenth centuries, the West was ready for them. Roman law and Christian theology had laid a foundation for believing that humans were rational beings, that there was considerable value in public discourse. Such discourse had already led to the highly rational and scholarly disciplines of jurisprudence and theology. Furthermore, the West had inherited a religion, Christianity, that was (at least initially) actively persecuted by the state. As a result, the church fought for and eventually obtained its own rights separate from those of the state. Merchants also secured many separate rights. Within this intellectual atmosphere, the universities of Europe began, scholarly guilds that had the right (as did other guilds) to conduct their affairs as they saw fit without undue interference from the state or established religion.

By 1200, two of the earliest universities, Oxford and Paris, based much of their curricula on Greek science. A statute, enacted in 1255 by the entire Faculty of Arts at Paris directed that Aristotle's natural science books be read by all students, even specifying the time to be spent reading them. Needless to say, all students were examined on their knowledge of Aristotelian logic and science. Not surprisingly, the view soon prevailed in the West that the universe was a rationally ordered system understandable by humans. Many believed that rational discourse among competing ideas would lead to greater understanding. This discourse, begun by the Greeks two thousand years earlier, was renewed in the West, as a growing body of "disputed questions" in physics, astronomy, and other disciplines were earnestly and openly discussed. Scholarly reviews of these questions and their various answers were prepared by the Masters and read by their students. Questions such as does the Earth turn on its axis? Is there a vacuum? Can things happen by chance? These were the hot topics of the day.

The states not only allowed, but occasionally even supported institutions where free enquiry was encouraged and scholars were chartered to discover the most consistent and theoretically powerful explanatory systems. This was so even though it was realized that the fruits of such quests would not always be to the liking of the state or church. Although individual states would at times renege, squelching academic freedom and scientific enquiry, the West as a whole was committed to the freedom of thought for the same reason it was committed to mercantile freedom: competition.

Modern science was the result, to no small extent, of the efforts of two men near the beginning of the seventeenth century. They were the Italian mathematician, Galileo, and the English barrister, Francis Bacon. Unlike Greek scientists, who two thousand years earlier had shunned hands-on experimentation, Galileo was all for empirical testing. Aristotle had claimed, on deductive grounds, that heavy objects fell faster than lighter ones. Galileo, through the simple expedient of trying it out (tradition has it by dropping two balls from the Leaning Tower of Pisa), overturned Aristotle's erroneous deduction. In so doing he brought into question all purely deductive or logical-seeming explanations not backed by actual experimental results. A mathematician, Galileo looked for and found a mathematical order underlying his empirical observations.

Hearing rumors that lenses could make distant objects appear closer, Galileo built the first telescope. Using it to observe stars, our moon, and especially the planet Jupiter and its four major moons, Galileo provided convincing support for Copernicus' theory that the Earth traveled about the Sun, not visa versa. Within three months, his results were published (at his own expense) as the book, *The Sidereal Messenger*. Galileo, not wishing to be scooped, had some five hundred copies delivered to influential figures throughout Europe. After the horse had left the barn, the Church put the aging Galileo on trial for heresy, making him recant that the Earth moved about the Sun. Legend has it that as they pronounced sentence on him he unrepentantly mumbled under his breath, "Yet it moves."

Protestant Northern Europe might also have dealt science an early blow similar to the Catholic south had it not been for Francis Bacon. Not a scientist, but an experienced barrister and vocal publicist, he championed the idea that science -- by reading the book of nature -- glorified God. God had intended that we should have dominion over the planet and science could provide us the power to take our rightful place. Through knowledge -- built up via science's methodical investigations of nature -- we could progress towards a new Golden Age. Far from being sacrilegious or heretical, Bacon assured everyone, science served God, revealed His Truth, and allowed us to help Him achieve His ends.

The only effect the Church's action against Galileo had was to place a chill on Italian science (and to launch the Church on an anti-scientific battle it could never win). Isaac Newton was born the year Galileo died. Francis Bacon's championing of science led directly to the founding of the Royal Society. In competitive Europe, science moved on. The genie was out of the bottle. There could be no turning back for the West or the planet.

The Greeks attempted to obtain total certitude by way of pure logic and reason. As this could be done without dirtying one's hands with lower-class practicalities, such an approach nicely matched their elitist attitudes. But, as Galileo demonstrated with his falling weights, logic without empirical verification could lead one astray, could produce unreliable, false information. In their search for absolute, foolproof knowledge, the Greeks latched onto logic as their road to absolute truth, but in insisting on certainty, they missed most of reality. Modern science, like Greek science, still insists on logical, often mathematical consistency, but its primary criterion for reliability is empirical verification, usually via clever demonstrations or experiments that show that the results claimed are likely to be the actual case. These demonstrations are then independently repeated by others. Only after such replications are the results accepted by the scientific community as reliable. Unlike the Greeks, no claims are made as to absolute certainty, only that what was is presented is the best, most reliable information we knew at this point in time.

The genius of this system was that it allowed reliable information and understandings to accumulate without everyone having to prove everything again from scratch for themselves. Because the results of other scientists could generally be counted on to be reliable, one could concentrate one's own efforts -- often for an entire lifetime -- on one small piece of the whole. There could be almost as many specialties as there were scientists.

For modern science to work its collective magic, all participants needed to adhere to rather strict definitions as to what did and did not constitute reliable knowledge. Appeals to religious authority, spiritual revelation, widespread public belief, pure logic by itself, human comfort, and economic advantage, were all rejected out of hand as unreliable. Only the empirically verifiable and logically consistent were accepted, and then only provisionally, always open to being shot down by new experiments or replaced by better explanations.

To protect themselves from unreliable information contaminating their journals, scientists instituted a screening process (beginning with the Proceedings of the Royal Society) of peer review prior to publication. Recognized experts in a given area reviewed proposed papers. If they could not be brought up to prevailing scientific standards of reliability, they were rejected. Science is not about freedom of the press, it is about a collective enterprise for the accumulation of reliable information. "The West alone," as Nathan Rosenberg and L. E. Birdzell noted, "succeeded in getting a large number of scientists, specialized by different disciplines, to cooperate in creating an immense body of tested and organized knowledge whose reliability could be accepted by all scientists."

Science wasn't just accumulation of verifiable facts. It was also an accumulation of explanatory theories that tied these facts together. Theory building has always been a risky business in science. A theory, no matter well it explains a multitude of facts, can be shot down by a single new experiment that contradicts the theory. Theorists are a tiny minority in science. They must face a virtual army of experimentalists, each one anxious to be the first to shoot down a theory with some clever experiment that contradicts it. Fortunately there is another, even larger army of experimentalists who spend their lifetimes generating multitudes of new, esoteric, reliable facts, all grist for the mills of undaunted theorists whose job it is to weld them together into a larger scientific whole.

What constitutes good theory? What is scientific understanding? William of Occam, a fourteenth-century British philosopher, suggested that -- other things being equal -- the best theory was the simplest theory. Science has generally adapted "Occam's razor." Theories, even if not contradicted by known facts, are considered in poor taste if not empirically verifiable (unprovability is exalted in religion but rejected in science). Also in poor taste are after-the-fact adjustments of theories to fit new empirical data, especially if these adjustments appear contrived or *ad hoc* in nature. Good theories not only explain the facts their creators intended them to, but suggest new possibilities that, when explored by new experiments, they also explain.

Why are we humans able to scientifically comprehend nature? It is amazing that an animal has evolved that can grasp, at least in its basic essence, its own evolution and that of the entire universe. Of course the "goal" of life from its inception has been the accumulation of information useful for its own survival and propagation. As noted by the late philosopher of science, Karl Popper, "From amoeba to Einstein, the growth of knowledge is always the same: we try to solve our problems, and to obtain, by a process of elimination, something approaching adequacy in our tentative solutions." Modern science is such a powerful evolutionary process for generating and accumulating reliable information that we should not be surprised that in a mere four hundred years it has homed in on an overall understanding of nature.

Science has proceeded, like machines, as another evolutionary process similar to genetic evolution. Instead of genes competing, it is facts and theories. Instead of nature selecting which will survive, it is humans. As with biological organisms and their amazing fit to the environment, science evolved to fit scientific theories to the universe they attempt explain. Unlike biological evolution, which leads to increasing diversity, science evolves towards increasing unity, a progressive trend resulting from human intentionality. It is based on our insistence on comprehensiveness, on having the pieces fit together.

Unlike biological evolution, which proceeds at a glacial pace, science evolved with lightning speed. Scientific protocol requires that theories be reduced to internationally understandable written descriptions (often in mathematical notation), and published in journals available to the public at large. The first to publish -- not the first to discover -- receives the credit. By hastening disclosure, new theories are rapidly subjected, world wide, to the scrutiny of other specialists (as well as being available to all for further elaboration). It is this strong selection pressure that has resulted in science's rapid convergence on reality. As Paul Gross and Norman Levitt noted: "Reality is the overseer at one's shoulder, ready to rap one's knuckles or to spring the trap into which one has been led by overconfidence, or by a too-complacent reliance on mere surmise. Science succeeds precisely because it has accepted a bargain in which even the boldest imagination stands hostage to reality. Reality is the unrelenting angel with whom scientists have agreed to wrestle."

So, how did scientists discover who we are and how we came to be? They faced the same difficulty as their religious predecessors, namely no idea at all of the immense depth of time that stretched billions of years into the past, and few clues as to the many forms of life that preceded current life. Nevertheless, biologists set out with scientific zeal to impose order on life's prodigious variety. They grouped known plants and animals together based on a system for classifying and naming life forms devised by the Swedish naturalist, Carolus Lineaus. They began the planet-wide search for previously undescribed and unclassified life. As European ships and explorers spread over the planet, expedition naturalists shipped well preserved specimens of exotic plants and animals they encountered to avid stay-at-home collectors.

From an anatomical point of view, it soon became clear enough where we belonged. A chimpanzee, shipped from Angola to London in 1699 was dissected by Edward Tyson. Orangutans and other apes soon appeared in European zoos (and on the dissecting tables of anatomists). Although obviously physically similar, there was still no reason to suspect we were descended from apes. The entire notion of such descent made little sense in a world believed to be six thousand years old.

Meanwhile, geologists were similarly classifying rock formations. They noticed that matching layers of rock occurred at widely spread locations. Some of these layers were embedded with matching types of seashells. A few even had matching types of fossil bones, presumably from extinct animals that perished in the Great Flood mentioned in the Bible. They also observed that different types of layers, piled one on top the other, formed matching sequences, also over wildly different areas. Although often jumbled up and, occasionally, even in reverse order, there did seem to be some sort of underlying master sequence.

In the early nineteenth century, the English geologist, Charles Lyle, suggested that the geological strata were sedimentary layers deposited on ocean bottoms by the gradual process of erosion over immense periods of time. These layers had been compressed into rock, and then, somehow, raised into mountains and, in the process, had been extensively rearranged. Thus the present jumble. By observing how past sedimentary deposits were being currently laid down and noting the depth of past layers, one could, roughly, estimate the ages of the various layers in the master sequence and, overall, how long it had taken for the entire sequence to form, i.e. the age of the Earth. Far from being a comfortable 6 thousand years old, as expected, geologists concluded that the Earth was billions of years old, an immensity of time that staggered our imagination. The depth of time was an unexpected shock.

A copy of Lyle's revolutionary *Principles of Geology* was obtained, hot off the press, by the then twenty three year old English naturalist, Charles Darwin for his round-the-world voyage of exploration on the H.M.S. Beagle. Darwin served as both the ship's naturalist and the intellectual companion of the ship's Captain, Robert Fitzroy. As with naturalists before him, Darwin's main job was to obtain and preserve specimens for shipment back to England. Having just read Lyle's books, Darwin also became an avid observer of never-before-described geological formations in South America. He wrote accurate descriptions and knowledgeable interpretations of South American geology which he posted back to England, much to the delight of his geological friends. Darwin was also an avid collector of fossil bones, shipping them back to England by the crateful.

Before crossing the Pacific, the H.M.S. Beagle stopped in the Galapagos Archipelago, a group of seven islands laying some seven hundred miles west of South America, but only a few miles from each other. What Darwin observed on the Galapagos didn't make sense to his orderly mind. Instead of the normal variety of birds he observed on the South American mainland, there were only numerous varieties of rather similar finches. Besides filling the role normally reserved for finches, finches were also doing what parrots and other birds normally did. The bird niches were all occupied by just slightly different species of finches, some with longer beaks, some with shorter beaks and so on. Why, Darwin wondered, would God have used just one type of bird for so many different jobs? He later wrote, "One might fancy from an original paucity of birds in this archipelago, one species had been taken and modified for different ends." Then there were the iguanas. Besides occupying their normal lizardly niche, other slightly different species of iguanas were eating plants normally eaten by ungulates and so on. All very strange!

On returning to England, Darwin spent a number of years describing the specimens in his huge collection, farming entire sections of it out to specialists, and writing an adventure journal of his voyages on the Beagle that was widely read by scientists and non-scientists alike. As he worked and wrote he mulled over the finches and iguanas of the Galapagos. The most obvious explanation was that all varieties of finches were descended from some original pair or flock of finches that arrived from the mainland, perhaps shortly after the geological formation of the islands. But how did they split into different species, each adapted to a somewhat different niche?

Darwin, long a fancier and breeder of racing pigeons, knew about natural variety in the offspring of animals, and that, to some extent, such variety was heritable. The key that explained it all occurred to him when he read an essay by Thomas Malthus which suggested that while human populations expanded geometrically their food supply only expanded arithmetically. Thus, in the long run, there was bound to be more people than food to feed them. Darwin realized this was also true with animals, with all life in fact. In a stroke of true genius, Darwin realized that for each generation some would be "selected" to live and some to die. Unlike domesticated animals, where humans selected who would live and die, nature herself would do the selection, automatically choosing from the variety available, those which were most capable of surviving, finding mates, and reproducing themselves.

Through a process of proliferating offspring, heritable variety, and natural selection, Darwin surmised that the original finches and iguanas on the Galapagos had diversified into a growing variety of specialists, each evolving to more efficiently utilize the resources in its selected niche. From this it was a short step to suggest that this was case for all life.

Darwin argued his case with great skill and scientific objectivity in *The Origin of Species*, published in 1859. Not wishing to needlessly inflame religious sensibilities with respect to humanity, he only discussed animals. Others were not so reticent, however, and the evolution of humanity, both physically and socially, became a hotly argued topic virtually overnight. A century and a half later the argument continues unabated.

After the publication of *The Origin of Species*, evolutionary enthusiasts were quick to suggest that we had evolved from some common ancestor with the apes to modern, European-dominated civilization. With almost no hominid fossil evidence available (and with little understanding of or empathy for the few remaining hunter-gathering cultures), those quick to apply Darwin's evolutionary insight to humanity quite mistakenly latched on to existing human races as being the living representatives of the evolutionary steps humanity had taken in its ascent from apes to its apex, Western civilization. What these early anthropologists and social historians didn't realize was that the common ancestor between humans and apes was actually many millions of years in the past, while the common ancestor for all current human races was only a couple of hundred thousand years in the past. Thus differences in races, far from being indicative of the long evolutionary history of hominids, only represented minor variations accrued since the recent origin of modern *Homo sapiens*.

This error of evolutionary interpretation, of drawing extensive conclusions without sufficient supporting data, had serious social consequences. It provided a scientific rationalization for the supposed more advanced races (i.e. white Europeans) to treat other races as less than human. From the mistreatment of slaves in the southern United States to the gassing of millions of Jews by Nazis during World War II, this mistaken notion of human evolution was used to justify the grossest of inhumanities.

In the early decades of the twentieth century, evidence gathered by anthropologists, such as Franz Boas, suggested that so-called primitive societies and peoples were, in actuality, not simpler, merely different. Languages and social customs appeared to be equally complex across all human societies. Although civilizations were more complex, they were a very recent phenomena, clearly unrelated to human physical evolution.

Meanwhile, an increasing number of the rare, hominid fossils were discovered. As physical anthropologists began to piece together the real story of human evolution it became increasingly apparent that our common ancestor with the apes was millions of years in the past while the current races of humanity had only recently diverged from each other. The damage had already been done, however. Many social anthropologists declared -- with some justification -- their independence from evolution and biology. Human behavior was molded by human cultures, they declared, and human cultures could be whatever humans chose them to be. Evolutionary thinking about human behavior and social organization was not only futile, but would encourage racism. Thus it should be strongly discouraged.

Removing human behavioral and social evolution from the scientific agenda created a problem for physical anthropologists such as Louis Leakey, however. Physical anthropologists wanted to know why hominids had physically evolved the way the fossil record indicated. They suspected that changes in behavior and social organization were key evolutionary forces. Just how had behavior evolved from our common ancestor with the apes to modern *Homo sapiens*? Leakey, with considerable insight, realized that we didn't really understand the behavior of our nearest relatives, the great apes, that our low opinion of their behavioral capabilities was based on observations made of captive animals in zoos or laboratories. Leakey surmised that ape behavior in its natural setting would be very different, perhaps even surprisingly human.

Leakey followed up on his hunch by selecting, encouraging, and even partially financing three young women -- Jane Goodall, Diane Fossey, and Buruti Galdikas -- to study, respectively, chimpanzees, gorillas, and orangutans. Leakey reasoned that females would pose less of a challenge to the male-dominated ape societies, would make more patient, long-lasting observers, and that fresh young minds not yet overly contaminated with the prevalent, negative scientific notions of ape mentalities, would be more open to the realities of ape behavior in the wild. Leakey could not have made better choices. Not only was our understanding of ape behavior revolutionized, but almost every human trait previously thought to be unique was brought into question. Slowly but surely it again became permissible to consider human behavior and social organization in evolutionary terms.

History is an academic discipline so entirely human, so tilted to the present (just the last five thousand years or so), that it is generally not considered science at all. Nevertheless, within history the scientific, evolutionary viewpoint has had most startling impact. Many historians were, quite legitimately, concerned only with the short run. They were satisfied with a purely descriptive approach. But the world historians who wished to cover the whole of the human civilized experience probed beyond the short-term appearance of randomness in search of longer-term trends or regularities. World historians looked for patterns, for lawful behavior.

Spengler, Toynbee, Sorokin and other world historians, writing in the early twentieth century, saw a cyclic behavior in the rise and fall of civilizations. They discerned a repeating pattern, not a long-term trend. Recently, however, world historians such as William McNeill, Alfred Crosby, and Clive Ponting, have placed human history into a scientific, biological, even ecological context. Their work has revealed a long-term trend that stands out over both the medium-term cycles of civilization and the short-term accidents of history. McNeill pictured the last ten thousand years as a working out of the implications inherent in our revolutionary adoption of agriculture. Crosby's biological interests are clear from his statement, "I insist that that which enables human beings to stay alive and reproduce and that which dispatches us to our eternal reward is worthy of our attention." This new breed of world historians no longer views history as a series of accidents or repeating cycles of rise and fall, but as a natural, evolutionary process whose major features proceed in a logical and scientifically explicable way.

Almost a century and a half after *The Origin of Species*, the evolutionary view provides a unity crossing all the major scientific disciplines concerned with explaining who we are and how we came to be. But the same rapid evolutionary process has also summoned forth other, darker understandings.

The predawn darkness of the New Mexico desert ended suddenly on July 16, 1945 as Robert Oppenheimer, quoted from the Bhagavad-Gita:

> If the radiance of a thousand suns
> were to burst into the sky
> that would be like
> the splendor of the might one.

As the fiery mushroom cloud rose above the brilliantly illuminated desert, he quoted from the Bhagavad-Gita again:

Now I am become Death
The destroyer of worlds.

Three weeks later, Hiroshima was incinerated in a blinding flash. Science's age of innocence was at an abrupt end. The atomic bomb was based on the purest, most advanced of the sciences, physics. Further, the bomb's development had been called for by the most respected of scientists, Albert Einstein. Looking back from the post-war vantage point of 1947, Oppenheimer noted that, "In some sort of crude sense which no vulgarity, no humor, no overstatement can quite extinguish, the physicists have known sin; and this is a knowledge which they cannot lose."

Francis Bacon's dream of science as being entirely of positive benefit to mankind had suddenly taken a bizarre turn. Scientists' curiosity led them into an area of knowledge humanity was ill-equipped to handle. But having pried open Pandora's box, they seemed powerless to select what emerged. They had released forbidden knowledge. It was now lose in the world, and there was no way it could ever be stuffed back into the box nor, apparently, the lid slammed shut against further evils.

Nor were the scientific creatures flying forth from Pandora's Box entirely nuclear. Rachel Carlson's *Silent Spring* alerted humanity to the ill effects of the artificial compounds being devised by organic chemists. These organic molecules had never appeared before in all the billions of years that life had evolved on the planet. Living beings had evolved no defenses against them, no means for neutralizing them or breaking them down. As increasing numbers of these unnatural organic compounds spread throughout Earth's ecosystems, the widespread environmental damage they could cause in even trace amounts became obvious.

It was clear that one species of life, *Homo sapiens*, its attendant machines and, especially, its rapidly snowballing store of reliable information and understanding accumulated by science, was becoming a pervasive force on the planet. One of the youngest sciences, ecology, had, since the turn of the century, made rapid progress in understanding how life as a whole, and the planet that supported it, interacted as a system. Ecology, an empirically-based science, measured and analyzed how energy, initially supplied by the sun, worked its way up the food chain from photosynthetic plants to top predators. Ecology also considered how scarce but vital constituents of life, such as carbon and nitrogen, were continuously circulated and reused in a closed, planetary ecosystem.

It is not surprising that ecologists were in the forefront of those suggesting that the effects of scientific research and technological progress might be costly. We could seriously damage the ecosystem, triggering not only a mass extinction of other life, but perhaps even threatening our own continued existence. Our modern, high-tech civilizations had become so dominant over nature, so all-encompassing, that we seemed to have forgotten that we were still dependent on the planet for basic ecosystem services.

In reminding us of these elementary facts, ecologists emerged in popular consciousness as guardians of life, the protectors of a fragile planet against the destructiveness released by malevolent physicists and chemists (not to mention the rapidly proliferating machines). Good scientists against bad scientists. Mesmerized by the power of science, but appalled by its results, young politicians latched onto ecology as the basis for a popular political movement; environmentalism. Fighting fire with fire, they turned science against science. Nothing new. Scientists have always reveled in upsetting apple carts, especially their own.

Not only has the power of science for both good and evil been evidenced in the physical world, but also in mental and spiritual realms. Science is about ideas, often powerful ideas. Its immodest goal has always been to reach an understanding of reality, of truth. It is not surprising that science has eroded many ancient authorities and traditions, beginning with those existing in Western Europe where science began. Some view this as a process of cultural demolition, of bullying science overwhelming defenseless local cultures, reducing them to quaint museum pieces or driving them to extinction. Others see science as one piece in the latest installment of what began with agriculture and, especially, civilization. In this view, cultural diversity peaked about ten thousand years ago.

Since then, the number of languages, of unique local cosmologies and religions has continuously declined. In this view, Christianity, Islam, and other great religions played their part in reducing cultural diversity by extinguishing local religions (thou shalt have no graven idols before thee). Still, science somehow seems especially pernicious (not to mention Hollywood).

In some democratic societies, particularly the United States, an increasing number of people are proclaiming their right to reject science if they choose, particularly the implications of evolution regarding humanity. Long accustomed to a separation between church and state, they now clamor for a separation between science and state. Failing (so far) to eject evolutionary teachings from public schools, many are teaching their children in private schools, even in their own homes.

Most people, scientists included, believe that science is one way of knowing among many. Both art and science are engaged in discerning truths about ourselves and our world. Art addresses the inner, subjective, private world of emotions and feelings. Science addresses the outer, objective, public world of physical forces and events. Music, which can exist independent of language, has little overlap with language-dependent science. Art and science peacefully coexist.

Does religion provide an independent, non-overlapping path to knowledge, to human wisdom and truth? Does science provide us with truth without human meaning? In undercutting religions by showing their cosmologies to be scientifically untenable, has science also destroyed, without replacement, the human meaning of religion? Three possibilities come to mind.

The first is that science may (eventually) provide human meaning. As we learn more about how great apes live in the wild and understand how we evolved our own peculiar capabilities and social arrangements, we may develop an increasingly sound, scientific basis for ethics and morality, eventually supplanting religions with science. Science might become a new, universal religion that will replace earlier superstitious beliefs as it unites humanity in a single, harmonious culture.

The second possibility is that as science overcomes its youthful reductionist and mechanical biases, as it develops an evolutionary grasp of the whole of reality, there will, as cosmologist Brian Swimme suggests, be an integration of "science's understanding of the universe with more ancient intuitions concerning the meaning and destiny of the human." In other words, a new religion will emerge that encompasses science yet retains the religious meanings that have always been beyond science.

The third possibility is that science will come to an end -- at least the sort of grandiose, big-picture science that has upset our religious perceptions of the universe. Perhaps the major ideas of science have now all been uncovered via an evolutionary process of discovery and exploration similar to the discovery and exploration of a new continent. The similarity is that a new continent is only discovered once. Once its major features have been mapped, the principle mountain ranges and rivers traversed and named, the age of discovery and exploration comes to an end. It is, as author John Horgan put it, "like the discovery of America -- you only discover it once."

Once the major scientific discoveries have been made, once all the big surprises like the size of the universe, the age of Earth, and the evolution of life, have been uncovered, once science settles down to increasingly finer details, the aging big discoveries will become old hat, yesterday's news. Newly uncovered details, the low hills and little streams, although they will continue, will not upset our world view. Religions adjusted to the Earth being round. They will soon adjust to the rest of science's Earth-shaking revelations. When science ceases to come up with major new findings, then religions will no longer be hostages to science, free to go off in a thousand different directions depending on the desires of their adherents. Human diversity, reversing itself again, will increase.

Human imagination always exceeded the constraints of objective reality.

For four billion years life never bothered to ask how it came to be. Then, becoming aware of its own mortality, it popped the question. Our socially-evolved brains supplied humanly-pleasing but dead wrong answers until, just four hundred years ago, science began its successful search for correct answers. Now we know how we came to be. We also know the answers to questions we wish we had never asked. As we look to the future, we are grateful for the good life that technology and science provides us, grateful to the machines that do most of our real work. But we are wary, even fearful, of who we have become, of the unintended consequences of our newfound power and knowledge. Having eaten from science's tree of knowledge, we have truly left the Garden of Eden.

FURTHER READING

Bowler, Peter J., 1988, The Non-Darwinian Revolution: Reinterpreting a Legend.

Desmond, Adrian, and James Moore, 1991, *Darwin: The Life of a Tormented Evolutionist.*

Dunbar, Robin, 1995, *The Trouble with Science.*

Eiseley, Loren, 1958, *Darwin's Century: Evolution and the Men Who Discovered It.*

Greene, John C., 1959, *The Death of Adam: Evolution and Its Impact On Western Thought.*

Gross, Paul R., and Norman Levitt, 1994, *Higher Superstition: The Academic Left and Its Quarrels With Science.*

Horgan, John, 1996, *The End of Science: Facing the Limits of Knowledge in the Twilight of the Scientific Age.*

Huff, Toby E., 1993, *The Rise of Early Modern Science: Islam China, and the West.*

Kitcher, Philip, 1993, *The Advancement of Science: Science Without Legend, Objectivity Without Illusions.*

Nitecki, Matthew H., and Doris V. Nitecki, eds., 1992, *History and Evolution.*

Newton-Smith, W. H., 1981, *The Rationality of Science.*

Rajecki, D. W. (ed.), 1983, *Comparing Behavior: Studying Man Studying Animals.*

Shattuck, Roger, 1996, *Forbidden Knowledge: From Prometheus to Pornography.*

Stevenson, Leslie and Henry Byerly, 1995, *The Many Faces of Science: An Introduction to Scientists, Values, and Society.*

Toulmin, Stephen, and June Goodfield, 1965, *The Discovery of Time.*

Wolpert, Lewis, 1992, *The Unnatural Nature of Science.*

III. WHAT IS OUR FATE?
Four finales

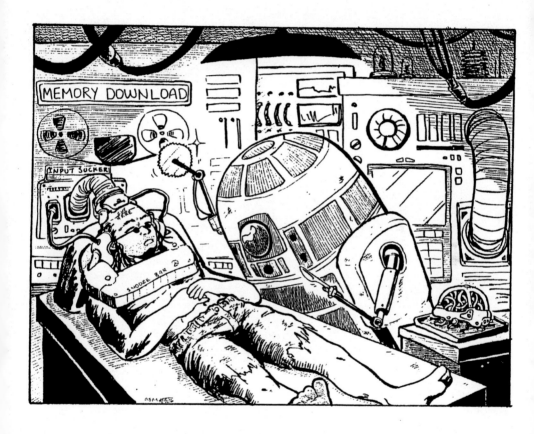

"Goodbye last human. Your memories are safely downloaded. You'll love your Positronic brain. Welcome to eternity."

CHAPTER 9

BOOM AND BUST
Humanity's Extinction Liberates Earth

What is our fate? Can science give us an idea of where we are headed and, if we don't like the direction, provide us with suggestions as to how we might change course? As our most objective, reliable, and extensive source of information, we feel that science should tell us our fate and how we might influence it. After all, science is predicting the fate of the universe and has an understanding of how four billion years of evolution has produced a diversity of species in the millions. One would hope that science could forecast the fate of a single species on one little planet. This appears not to be the case, however. While good at making long-term predictions for simple systems when forces are few and historical precedents numerous, science has difficulty making long-term predictions for complex, one-time situations influenced by many parameters, some of which are random. It may not, even in theory, be possible to make such predictions. Small effects may always send complex systems off in entirely unpredictable directions.

Short term predictions are another matter. A growing number of futurists are paid by governments and businesses to make short-term forecasts, typically just ten or twenty years into the future. While not always right, these forecasts are good enough to be useful for planning. A new academic discipline has emerged to train and guide these futurists. What they do best is extrapolate current trends in an objective, often mathematical manner. Many short term predictions are of a contingent nature, i.e. if such and such happens, then it follows, with some probability, that the outcome will be so and so. Such predictions are useful in choosing between alternative courses of action.

Considering the fate of humanity, our primary interest is not in the distant future, millions or billions of years from now. We realize that such predictions are risky and relegate such speculations, interesting as they may be, to the realm of science fiction. On the other hand, the futurist's modest projections of current trends a decade or two into the future is not what we are looking for either. What we want to know is the probable outcome of what appears to be a likely collision between an expanding and aggressive humanity and a finite planet. Considering the speed with which humanity is approaching planetary limits, it seems humanity's moment of truth will be upon us and resolved, one way or another, in the next century, perhaps sooner.

The combination of increased agricultural yield and reduced death rates created an explosion of human population. Humanity is now doubling every 40 years. This rate simply cannot continue, because we are already using 40% of the primary productivity of the planet. That is, 40% of the plants, the primary producers on this planet, are already used one way or another, by humans. The simplest arithmetic suggests that restraints will be placed on the growth of human civilizations. If scientific or technological breakthroughs are going to save us, they will have to appear soon. If, to avoid a showdown, we are to suddenly restrain ourselves, we will have to do that soon. No matter what happens, we should be beyond our moment of maximal crisis by the year 2100.

Interestingly, we are not slowing down as we approach our maximum pucker point but are, quite perversely, rapidly accelerating. As Lester Brown of the WorldWatch Institute pointed out, until recently the growth in population during any individual's lifetime was hardly noticeable, but since mid-century "the pace of change has been breathtaking." Older folks alive today (born prior to 1950) have seen more population growth within their lifetimes than has occurred over all previous generations since our ancestors parted ways with the apes.

Although big game hunting and, especially, agriculture launched significant population growth, it was the combination of science and machines that gave it its final, suicidal burst of speed. Full throttle into the wall! Science's gifts of sanitation, antibiotics, immunizations, and insecticides cheaply and effectively prolonged human life. Technology's gift of machines cleared and leveled the land, pumped water, and transported food about the planet. Since 1950, consumption of seafood has expanded by a factor of four, as has fossil fuels consumption. In the single decade from 1985-1995, the WorldWatch Institute estimates that the planet's economy expanded by $4 trillion, more than the total economic expansion of all civilizations from the Sumerians to 1950!

It is hardly surprising that we are hitting up against planetary limits. According to the WorldWatch Institute (again), grain land area peaked in 1981 when land lost to salination, desertification, and urbanization exceeded new land brought into grain cultivation. Planet-wide use of fertilizers peaked in 1989 because increased use would not have resulted in increased productivity. Total grain production peaked in the following year, 1990. The amount of usable irrigation water also peaked in 1990. *Homo sapiens*, after less than 200 thousand years of expanding population, a blink of the eye in evolutionary time, is encountering planetary limits. True biological success!

We now consume resources faster than they can be naturally replaced. We started down this path as well-organized big game hunters. Soon most of the big game was gone, followed by smaller game, then by wild fruits and vegetables. As agriculture expanded, more land was grazed or put to the plow. Increasing portions of agricultural land were lost forever to soil erosion, salination, and desertification, all processes that began with Sumer and other early civilizations. Now there are no new continents, no new fertile places to farm. Food production has peaked while Earth's human population continues to soar. With the entire planet harnessed to feed a single species, it appears that we are about to run short of food. Our moment of truth approaches.

While predicting humanity's future is beyond science, we can, as futurists often do, consider several alternatives. By changing various key assumptions we can consider how these changes might lead to starkly different futures. Scientists, for instance, disagree over how robust the planet's ecosystem is. Some believe it is like a highly interconnected house of cards, and that we could easily trigger its sudden collapse. Others believe that living associations are loosely connected, and mere humans are not about to cause planetary ecosystem collapse. Some believe humanity has control over its own destiny, others feel that this is wishful thinking -- we are not in control and never have been.

One could generate a list of differing views within science that could lead to an infinity of possible futures. I outline, in what follows, four alternative futures that represent a fair cross section of future possibilities. My intention is not so much to predict the future as to explore the implications of variations in today's scientific viewpoints. What are the ways humanity's penultimate crisis might be resolved?

Selected to stress differences, these alternatives are in considerable disagreement with each other. As each viewpoint is described, I attempt to be a proponent of that viewpoint, to make the strongest possible case for it. Thus the following chapters sharply and intentionally disagree with each other. The four futures are boom and bust, robotic triumph, pigship earth, and chimpanzee paradise. Readers may decide for themselves which alternative is most likely and which most desirable.

The first future is boom and bust. To add a touch of further variety, two bust variants will be considered. The first is based on the simple observation that many civilizations have gone bust in the past. These were "adjustments" of overly complex or populous civilizations to simpler, more efficient societies with smaller, more sustainable populations. Analogously, the present global-wide, interconnected, highly-complex and carefully integrated international system supporting 5.8 billion humans would be replaced by a simpler, more fragmented systems that would support a smaller number of humans. Complex societies have always been difficult to maintain, vulnerable to many problems of their own making, and hence prone to collapse. By considering the collapses of earlier societies, we can make reasonable projections about the impending collapse of our own. In this I shall follow the work of Joseph A. Tainter, as suggested in his *Collapse of Complex Societies*.

For simple societies, increases in complexity can be beneficial. We saw earlier that complex insect societies, the insect superorganisms, pushed solitary insects to the periphery, leaving them just a few crumbs, so to speak. We have seen how civilizations have also pushed less complex societies, the hunter-gatherers, to the periphery, leaving them even fewer crumbs.

As complexity increases and society grows, payoffs from increased complexity and greater size diminish. The easy pickings are gleaned off first. As a result, further gains become increasingly difficult. At some point, the advantages of any additional complexity, of larger organizational size, are offset by the diminished size of the returns. The growth of complexity doesn't stop at the optimal point, however, due to the jury-rigged nature of evolution itself. Nature never redesigns things from scratch. Improvements are tacked onto existing structures. Civilizations are no different. New problems are solved by adding more specialists, another layer of bureaucracy some technical fix. As the number of problems civilizations encounter are without end, so to is the growth of their complexity.

Ant superorganisms are fortunate that their growth was restrained by other life. Checks and balances coevolved while they were slowly exploring the possibilities of the superorganism realm over millions of years. This is why ants have been sustainable and successful for so long. In their heart of hearts, however, ants are just as greedy as we are. They were saved from themselves, restrained by others in a graceful manner.

Human superorganisms are less fortunate. Without effective checks from other life, our civilizations zoomed right past the point of optimality into heavy diminishing returns. Like earthquakes, they built up complexity tensions, then released them, all at once, in single spasmodic adjustments back to simpler societies that were a mere vestige of grander pasts.

Rome is the classic case. Initially the Roman Empire, with modest investment in military and transport, brought in immense surpluses from all the shores of the Mediterranean. In this, Rome reaped an outstanding return on a small investment. As time passed, however, the empire became ever more bureaucratic, taxes rose, standing garrisons grew ever larger. In the increasingly disgruntled provinces, elites were paid more for doing less, and, increasingly, oppressed home-front masses had to be bribed with ever larger doles and bloodier circuses. Eventually, the reserve capacity to respond to new problems was exhausted. Rome collapsed.

The Maya are another classic case. As surpluses and complexity grew, competition increased among the Maya city-states. An arms race spiraled out of control. No city-state dared opt out of the race, however, for to do so would have caused it to be instantly been grabbed by a neighbor. The fragile jungle environment was eventually exhausted. The city-states, locked in mortal combat, collapsed together, and life reverted to a simpler, pre-civilized state as the jungles reclaimed the once proud monuments.

Our situation today, while similar to these true stories of doom, is unique in several ways. Unlike earlier times, the world today is teeming with complex societies. Somewhat like the Maya dilemma, each expression of the human superorganism is entangled with the rest. As a result, collapse, when it comes, may be global, not local.

It is in the interest of the major powers that the current world system not collapse in a spasmodic human earthquake that devastates us all. As the industrial countries still have considerable wealth, it seems likely that they will be able to carry on for some time longer. Thus collapse is not an immediate threat, although it could occur at any time. The longer it is staved off, however, the more spectacular it will be when it finally comes. Are we really headed towards a collapse?

The signs of diminishing returns past the point of optimality are everywhere. It is easy to see the signs of runaway consumption. More farmland is lost every year than is gained by clearing the last forests on the planet. Food production in the most populous country, China, peaked in 1990 while its population continues to grow. Despite all efforts, China's food supply cannot keep pace. The result, almost certainly, will be the increased reliance on imports and an increased scarcity (and hence increased price) of basic food commodities worldwide.

Oil, on which the current high yields of agriculture are based (and with which vast quantities of food are redistributed about the planet), is rapidly being used up. Oil fields are being depleted faster than replacements can be found. All the easy oil has already been tapped. Other vital minerals are following this same downward curve. As less rich ores are tapped, greater amounts of energy must be consumed for refining, and larger mounds of waste are left behind. Nor are diminishing returns in any way limited to the physical. Society itself has hit diminishing returns. Governments are becoming incapable of handling worsening situations. Under these circumstances, even the most brilliant and well-meaning politicians appear inept (even more so than usual).

These obvious signs of a complex, worldwide social system heading for collapse have not been lost on a well-informed citizenry. People everywhere are concerned with the pending collapse of society. At least some segments within society are searching for graceful ways of moving in the direction of decreased complexity. Instances of this search are evidenced in lifestyles that do not depend, primarily, on industrially produced and transported resources. A grow-your-own, make-your-own sort of life is gaining adherents. More extreme are the growing number of survivalists who, certain of collapse, are stocking up on life's basic necessities, hoarding ammunition, and training themselves in the use of weapons and military tactics.

When it comes, the collapse might be considered a normal readjustment to a less complex, less populous, more efficient way of life. A corrective adjustment, if you will. Perhaps such a collapse should be welcomed -- the sooner it happens the better. The longer this necessary adjustment is staved off, the greater the resource consumption and waste production required to support an increasingly inefficient and larger international society. The longer the crash is put off, the more the planet will bear the scars of civilization. With a cold-hearted logic, we can conclude that taking our medicine now instead of later might be a good thing.

From the viewpoint of the few hunter-gatherers hanging on in the Amazon rain forest, a quick, clean global collapse of the industrial economies would be a blessing. With gasoline and diesel fuel unavailable, chain saws and bulldozers would fall silent. Without ships to transport wood veneers and beef to the industrialized countries, the invasion of the Amazon would stop.

For those living in large cities, however, a global collapse would be an unmitigated disaster. They have become dependent on food grown elsewhere, often halfway around the planet, food that requires a high level of organization and energy for its growth, transport, processing, and distribution. If a global collapse disrupts oil supplies, or if oil supplies become scarce and hence expensive, then the current world population (let alone an even larger population) would not be supportable. Mass starvation and brutal conflict over the scraps would soon take it down to a level that was supportable -- locally supportable without the benefit of fossil-fuel machines. Without appropriately distributed oil and functioning fossil-fuel machines, the world's supportable population is probably a billion or less. A sudden collapse could reduce it even further. While many millions would remain after the collapse, most of the billions would perish.

This would not be the end of civilization. Survivors, sadder but wiser, would pick up the pieces and move on, perhaps merely to boom and bust again as other animal species have repeatedly done. Perhaps the coming bust will be much more than an "adjustment" of social complexity and population, however. Perhaps it will be a permanent change, one brought on by irreversible effects of agriculture, civilizations, machines, and science. Rather than driving around the block in another cycle of boom and bust, perhaps we are on a one way street. It may be a dead-end street for civilizations, even humanity itself.

From a biological perspective, the human superorganisms we call civilizations are Johnny-come-latelies, mere biological infants. Human superorganisms have already achieved biological noteworthiness, however, thanks to the spectacular increase of their individual members from five million to five billion in only ten thousand years. This rapid growth has given human superorganisms a biomass equivalent to all ten thousand species of ants combined. Human civilizations are biologically noteworthy among superorganisms in being the only ones to have evolved culturally instead of genetically. In this we are a never-tried-before evolutionary experiment.

This cultural evolutionary experiment of human superorganisms has been spectacularly successful -- if one equates success with biomass. The very speed and magnitude of success are biological red flags, however, for they suggest that human civilizations have proliferated without effective restraint by other life. If human superorganisms follow the usual biological cycle of unrestrained growth, there is no question of the outcome; the unrestrained will be restrained. Resource scarcity, self-pollution, or restraint by other life will put a halt to the expansion of human flesh. There is, biologically speaking, nothing unusual about this scenario. Life is always slipping past normal restraints, even if only locally and temporarily. The boom is always followed by an even faster bust. We presume that human superorganisms will follow this biological cycle but on a planetary scale.

Algae in cold-weather ponds blooms every spring, right after the spring thaw. Due to their rapid multiplication, the algae soon consume all the nutrients which have accumulated over the winter and their population crashes. The next winter the nutrients accumulate again, and the cycle is repeated. In our case the planet is the pond. If we suddenly consume the planet's nutrients, it could be millions of years before they could regenerate themselves. This would not be just another civilization crashing in a long history of crashes.

The evolutionary fate of most lineages is dead-end extinction. Species of large mammals typically last about four million years. A few of them continue on by way of daughter species, but most terminate without issue. The evolutionary tree is a tangle of dead-end twigs and branches. Although mass extinctions occasionally exterminate species in wholesale lots, the end for most species is lonelier. Evolution has a penchant for mindlessly promoting the immediately useful without concern for the long haul. This short-sighted approach is followed by most species. They take advantage of some change, some new opportunity, some temporary gap in the ecological pathways. Genetically modifying themselves to milk it for all its worth, the transient opportunity usually goes away, leaving them high and dry with what is, in the end, a worthless, specialized, and usually fatal adaptation. That's the way of life.

By two million years ago, our ancestors became dependent on the cultural transmission of information between generations. This led, some ten thousand years ago to agriculture, and some five thousand years ago to civilizations. We will learn, in the next hundred years, whether or not civilizations are dead-end paths for our current, highly organized way of life. Cultural evolution, as is certainly the case for genetic evolution, promotes what is immediately useful, even if it is harmful, perhaps fatal in the long run. Consider agriculture: Farming does allow a larger number of humans to be supported in the short run. But if farming always and necessarily causes irreversible and increasing damage to the ecosystem, then in the long run the ecosystem may be able to support fewer humans. Agriculture could be a dead end.

More complex systems require greater energy for their maintenance. Civilizations require a continuous flow of energy, many times greater on a per capita basis than required to sustain hunter gatherers. Modern civilizations, with their fossil fuel machines, are energy intensive. They may represent another dead-end approach to life.

Would humanity ever allow itself to proceed down a one-way path to extinction? Many scientists believe that we may do just that. The primary reason may be that humans are not in control. Since the first city states of Sumer, human superorganisms have been competing both militarily and economically. The best organized and most efficient human superorganisms have survived and prospered at the expense of the less efficient. Starting a thousand years ago, the highly competitive and aggressive countries of Western Europe developed capitalism, which has now spread planet wide. Economic efficiency is the evolutionary force controlling humanity. God Efficiency is in charge. Like biological evolutionary forces, efficiency is blind to the future, lives for the present, and hence is irresponsible.

But aren't we taking control by becoming environmentally conscious? Or in reality are efforts to stem the rapid rise in human population, consumption, and environmental impact so small -- compared to the destructive tidal wave of humanity and its machines -- as to be insignificant? In this view, conservation moves to date are just palliatives thrown up by so-called democracies while the global corporations, which are in real control, continue to vie with each other without regard for protecting the environment, let alone reducing resource consumption. Where laws protecting the environment have been passed, corporations continue to pollute on a massive scale, preferring to pay the minuscule fines or simply move their operations to countries with laws more to their liking. The impact of environmental measures on the poorest countries, where the greatest increases in human population and environmental destruction are taking place, is essentially nil. In these countries the scramble is on to clear the last forests, to grow a bit of food, and to burn the last wood to cook it.

It is fashionable to believe that our current difficulties are the result of some spiritual deficiency, a defect in our Western psyche. If this defect could somehow be corrected and the masses converted to being faithful, peace-loving, green-friendly environmentalists, our impending collision with the planet could be avoided. Several decades of environmental preaching have had little effect. Industrialized countries, where consumption has skyrocketed, remain wedded to their shopping malls and the good life. Life has always strived to maximize its numbers, to grab the largest share of resources it can. Why should we expect to be different? We are doing what comes naturally, what evolution has fine-tuned and ingrained in all life.

We speed onwards to our doom, even accelerate. Increasingly aware of our fate, it is too late to regain control, the inertia of humanity, of the environment itself will propel us full speed into the wall. We will still be accelerating when the crash occurs full force. Like a slow motion nightmare, we will be aware of what is happening, but powerless to stop it. Exactly how the end will come is difficult to predict, so we must content ourselves with a few highlights.

The scale of industrial activities has become threatening. According to Lester Brown (WorldWatch Institute), industrial flows of nitrogen and sulfur are both larger than the natural flows. Human-induced circulations of cadmium, zinc, arsenic, mercury, nickel, and vanadium are twice natural flows. We might inadvertently switch the geophysical state of the Earth into some new, unfavorable, mode. Our carbon dioxide emissions could cause a runaway greenhouse effect that could melt icecaps, flood low-laying land, and drastically alter climate.

A nuclear war seems less likely now then it did a couple of decades ago, but what will happen when we run seriously short of oil? The fight over the last easy oil could easily escalate into all-out nuclear war.

Nor should we lightly dismiss the possibility of an infectious disease wiping out humanity. The Black Death killed a third of humanity. Our indiscriminate use of antibiotics has moved infectious agents of major diseases well along the pathway to total resistance. The combined biomass of humanity is the biggest one-species jackpot on the planet, and an astronomical horde of microbes are busily working to be the first to find the winning combination. It is only a matter of time.

Most threatening of all, however, may be the plows and chain saws as usual. When the forests are gone, when all arable land is intensively farmed, and when the human population doubles yet again, the planet's ecosystems may be permanently, irreversibly damaged. A planet devoted entirely to humans and their domesticated plants and animals may not be viable.

It is conceivable that, in our demise, humanity might lose the status of superorganism. Perhaps no civilization will remain intact. The surviving individuals may revert to more sustainable hunter-gatherer ways. If so, civilizations will have been a flash in the pan, another failed biological experiment, a temporary fluke in the continuing saga of a strange species.

Easter Island is the best preview of coming attractions. It displays, in miniature, many key ingredients of total collapse brought on by ecological neglect. Easter Island is isolated in the mid Pacific; a thousand miles from nowhere, a miniature Earth. A few Polynesian settlers arrived on Easter Island and, multiplying as life does when unrestrained, soon numbered in the thousands. Small as it was, there was room for two competing, complex societies. An expression of their competition was the carving of enormous stone statues erected on seacoasts far from the quarries.

Transporting these statues to their seaside vistas required they be rolled on logs. The trees on the island were cut to provide these statue-transport logs. As the last forests disappeared, do you think that these deeply religious societies stopped quarrying, carving, and transporting statues? Of course not. Competition for the most statues intensified as the last stands were cut. In the ensuing ecological disaster, most islanders either starved or were victims of vicious warfare-- it was not a time to be neighborly. The population, now numbering in the hundreds instead of thousands, never recovered. Easter Islanders were living in a semi-savage state when contacted by Europeans. The story of their own civilization's demise had been lost and had to be recovered by archaeologists.

If our bust is sufficiently severe and abrupt, it is conceivable that we might even become extinct, perhaps taking many other species down with us I the final catastrophe. Life on Earth has undergone mass extinction at least five times in the past. It is becoming clear that it is now well into its sixth mass extinction. All mass extinctions have been due to a loss of habitat. In the previous extinctions, this loss was due to abrupt climatic changes -- very abrupt in the case of the meteor which struck the Yucatan peninsula 65 million years ago with an explosive force 10 thousand times as great as all of humanity's nuclear weapons combined.

Another comet struck Earth 10 thousand years ago -- human agriculture. This mass extinction does appear to be unique in being triggered by life itself -- a single species at that.

As life all about us becomes extinct, it seems likely that we too will suffer the same fate. But perhaps not. Not only might we survive and might avoid a serious crash, we could prosper beyond all recognition. It has happened before in Earth's long history.

FURTHER READING

Asimov, Isaac, 1978, *A Choice of Catastrophes*.

Catton, William R., Jr., 1980, *Overshoot: The Ecological Basis of Revolutionary Change*.

Eldredge, Niles, 1995, *Dominion: Can Nature and Culture Co-Exist?*

Garrett, Laurie, 1994, *The Coming Plague: Newly Emerging Diseases In a World Out of Balance*.

Leakey, Richard, and Roger Lewin, 1995, *The Sixth Extinction: Patterns of Life and the Future of Humankind*.

Tainter, Joseph A., 1988, *The Collapse of Complex Societies*.

Wagar, W. Warren, 1991, *The Next Three Futures: Paradigms of Things to Come*.

CHAPTER 10

ROBOTIC TRIUMPH
Our Mind Children Inherit the Galaxy

Gloom and doom. The Old Testament prophets tell us that the world's imminent demise is just around the corner. Thomas Malthus gave scientific weight to the tradition of doom with his description of how Britain's population was increasing geometrically while its food supply was only increasing arithmetically. His conclusion? Massive starvation would soon set things right. Although Malthus' conjecture inspired Darwin's Theory of Evolution, its forecasts for Britain's demise were wrong. New resources, far from increasing only arithmetically, increased even faster than geometrically.

A few decades ago, scientists -- ecologists this time -- became concerned we might crash soon. Environmentalists were delighted to have the scientific support of ecologists in their fight to save the planet. But studying the matter further, the fickle professional ecologists concluded that nature, by way of massive volcanic eruptions, extensive droughts, widespread fires, and severe Ice Ages, has already been more disruptive than humanity could ever dream of being. Nor, they informed the turn-back-the-clock return-to-nature environmentalists, did there appear to be any single "natural state" for us to return to. Ecosystems were constantly changing, sometimes with considerable speed. The ecologist's earlier view of delicate, interlocking ecosystems, was replaced with a view that ecological associations in nature were robust. A crash did not appear likely after all.

What were environmentalists going to do? They had lost the heart of their scientific rationale. Not to worry, a few scientists still supported them. There was still some small chance that we might cause a collapse. Besides, the younger generation had already been converted. It had become an article of environmental faith that we were headed straight for the brick wall, that only a spiritual greening could save us. As in previous generations, however, there remained those rational minds (especially those who still had to work for a living) who continued to dismiss perennial forecasts of gloom and doom as the coffee-house wailing of the idle offspring of the rich -- existential cry babies.

While it may appear that we are pushing up against planetary limits, in actuality we are just experiencing a few minor growing pains as we transition the planet from genetic to cultural domination. We remain a long ways from the carrying capacity of Earth. The planet could support 10, even 20 billion people-- perhaps more. Technological and scientific progress have more than kept pace with population growth. As Francis Bacon predicted, the lot of the masses has dramatically improved, especially in societies that have had the good sense to actively embraced industrialization.

In industrial societies people are better nourished. Life expectancy is at its highest level and still increasing. The physical environment we live in is generally becoming cleaner, not dirtier, as technological advances allow us to recycle materials and safely dispose of industry's non-recyclable residue.

Far from coming to an end, economic growth can continue indefinitely. It is based not on an increasing consumption of scarce raw materials and energy as the doomsayers would have us believe, but on technological and scientific advances. Optical fibers made of inexhaustible silicon that carry millions of phone calls have replaced wires made of scarce copper which carried only a few conversations. Computers made of silicone chips and interconnected via optical fibers (the Internet) are replacing mail, journals, magazines, newspapers, and books. More information is being exchanged, yet fewer trees are needed for the paper, less oil for the airplanes and trucks.

Those alarmists who keep predicting that severe scarcity is just around the corner have been proven wrong, not because resources are inexhaustible, but because we keep finding cheaper and better ways of processing them. The substitutes that are better and cheaper than the original. Julian Simon, author of *The Ultimate Resource* and an articulate spokesperson for continued economic and industrial growth, bet Paul Erlich, an environmental alarm popularizer (*The Population Explosion*) that the price of copper would fall over the next five years. Simon easily won because the prices of most raw materials have steadily decreased for several hundred years. This widespread, favorable trend is due to hard-working machines, the bonanza of fossil fuel energy they eat, and our accumulation of massive amounts of reliable information via science. Understanding our reliance on fossil fuel machines, some alarmist suggested that when the oil for these machines runs low, the party will be over as oil prices skyrocket. Not so, counter the believers in continued progress. Energy itself has also become cheaper over time. This trend will continue as safe nuclear fission and, especially, efficient, clean, and inexhaustible solar power replace and make oil power obsolete. This will conserve remaining supplies of oil for use as feed stocks for the production of plastics and other organic chemicals.

As great as the benefits from science and technology have been, even these contributions will be dwarfed by the coming scientific benefits of the biotech revolution. No longer must we rely on crosses between the paltry number of species we can coax into interbreeding. Breaking down all genetic barriers, we will be able to transfer desirable traits from one species to another no matter how distantly related. We will be able to create plants that thrive in farmlands lost to salination or desertification. Plants will be designed to provide their own fertilizer (nitrogen fixing) and pesticides. Food production, as it enters an era of mass "factory" production, will become so efficient that the land devoted to agriculture will fall even as population climbs. Unlike the original agricultural revolution -- when weeds, vermin, and microbes (diseases) unfairly cashed in on civilization's artificial ecosystems alongside humans and their domesticates -- the biotech agricultural revolution will foil these opportunists, spreading genetic havoc among their ranks.

Taking the long view of life on earth, we might parse it into three eras. The first would be life before encoding information via DNA, an era when information and metabolism were one and the same. This first era lasted only a few hundred million years. The second is that of DNA, the era of genetic evolution and the selfish gene (not to mention cooperating genes). The second era has lasted four billion years. While it is still going strong, it is being seriously challenged. The third era is the cultural era, an era of information separate from metabolism, separate from DNA, even increasingly separate from DNA-based organisms.

This is not the first time that one form of life has forced other forms of life to adjust to its own, selfish agenda. Such biological takeovers do occur on occasion and are major turning points in the evolution of life. Photosynthetic life revamped the planet over a billion years ago when it tapped an immense new source of energy. It used solar energy to mine water for its hydrogen, releasing oxygen as waste. This was a case of one life form massively polluting the planet. Photosynthetic life established a new world order, forcing all other forms of life to adapt to the greatest pollution event of all time -- the wholesale release of oxygen into the atmosphere. Oxygen physically transformed the planet, literally rusting the oceans. Cultural life has tapped into a new energy source, fossil fuels. Again there are poisonous byproducts, although we will switch from fossil fuels to solar power or nuclear fusion long before we begin to match pollution caused by photosynthetic life.

In our case, there is some disruption of life on Earth, just as there was when photosynthetic life took over. But, as before, life will evolve to cope with the transformed planet we will assist it in so doing. With the advantage of hindsight, we can state that the planetary grab by photosynthetic life was a good thing, that its massive "pollution" proved, in the end, to be beneficial. Without oxygen and the high levels of energy release it enables, there would have been no animals. We should be grateful that some environmentally conservative bacteria didn't talk the first photosynthetic experimenter out of its new, "high-tech" process as being likely to despoil the environment. Who is to say that great good will not happen again? Out of the upheaval and destruction, out of the mass extinction of now archaic life forms, new life will arise that transcends it all. Perhaps machines, by utilizing vast, previously untouched stores of fossil fuels, are the new photosynthetic life, running out of control, changing the planet forever.

In any event, the cultural era of evolution on Earth is well under way. A new world order has been established. We humans and our domesticated plants and animals (not to mention our camp-following weeds, vermin, and diseases) are the winners. Other life forms are the losers. They need to adjust to us, not visa versa. We are in charge of evolution now. Except among the weeds and their microbial ilk, traditional Darwinian evolution is no longer of much significance on this planet. Agriculture has overwhelmed most of the natural world. Natural selection, for the most part, has been replaced with human selection. As the take-charge winners, we have become even more successful than our highly organized, hard-working predecessors, the ants.

Ants were rarely able to go beyond city states, colonies of about five million ants (with a combined weight of a cow). True, a few ant species were able to form empires of sorts, but such empires failed to take hold and remained rare curiosities. Not so with the human city states. Shortly after they formed, they coalesced into empires, super-superorganisms that competed at a higher level, entire sections of the planet competing against each other. Now, we are witnessing the emergence of the first planetary superorganism, a life form harnessing the entire planet for its own benefit. Sadly, winners imply losers; the planetary pie is only so large.

It is to easy to feel sorry for the losers. We naturally cheer for underdogs. Yet there are hard choices to be made. As humans increased, it was inevitable that other species would decrease. As we increasingly made their habitats our own, it followed that the variety of life would decrease. Natural ecosystems have now been replaced, in a wholesale manner, with human-dominated ecosystems. Life not adapted to the planet's new, dominant type of ecosystem is on the wane. How could it be otherwise? It is not clear that this is bad. Wouldn't the planet be as well off with one million species as with 10 million? Why not take the advice of Genesis 1:26;

Be fertile and increase, fill the Earth and master it.

In our victory we are magnanimously setting parks aside to preserve the outmoded losers, life that is unable to compete on its own in the New World Order.

Just as there are winners and looser species with respect to the New World

Order, there are winner and loser societies. Some societies are having difficulty in coming to grips with the modern world. In fact, they seem to be purposely shunning the New World Order. Clinging to corrupt or authoritative regimes, extolling religious fundamentalism, actively opposed to rationalism, capitalism, science, and freedom of the press, these countries appear to have purposely handicapped themselves just as competition was heating up planet wide. Because accelerating change is the hallmark of the modern era, countries that can handle change will prosper at the expense of those unable to do so. Evolution continues -- cultural evolution.

As the biotech revolution swings into high gear, food production will increasingly resembles industrial processes. Already sugar cane is being replaced by an entirely artificial chemical, isoglucose. As demand for old-style, agriculturally-grown food falls off, the economic position of low tech countries will worsen. Unless they get with the program in a hurry, they may have to be written off, allowed to collapse, and re-colonized at some later date. Civilized humanity, after all, replaced recalcitrant hunter-gatherer societies that refused to get with the program.

For those with eyes to see, the future has already begun. The planetary superorganism has arrived. Not a United Nations world government, not some utopian disarmament, not even an international Esperanto language, but while we weren't looking, a global, cosmopolitan culture has coalesced into being among the industrialized societies. Although this global culture's roots lie in the West, other regions have contributed. Eastern Asia, especially Japan, are increasingly influential.

National leaders no longer have the clout or significance they used to have. Power has moved to multi-national corporations and a multitude of decentralized but highly efficient decision makers interconnected by rapid global communications. The world economy is in charge. God Efficiency reigns supreme. The ants, if they could only speak to us, would approve!

Those countries and regions that are doing well in the new world system have the accumulated knowledge of humanity at their fingertips, a well-educated work force to access it (including large numbers of engineers and other professionals), and the capital, financial structures, and entrepreneurs to pull it all together. Japan is the epitome of the new order. Her masses of engineers and workers pride themselves on efficient production of the world's best-designed products. Dedicated to saving for their own individual futures, the Japanese masses have contributed the capital needed for their country's future. The United States, on the other hand, appears to be more interested in short-term profits than long-term investments, in buying as opposed to saving. Lawyers outnumber engineers. Although still the planet's most adept and creative researchers, scientists in the United States are increasingly financed by Japanese firms.

Robots were an American invention. Once 50 U.S. firms manufactured robots. Now not a single independent American firm remains. With only 2.5% of the world's population, Japan has 70% of its robots -- almost 200 thousand of them. Robots have all the virtues of earlier machines and more. Since human operators are not required, heating and air conditioning can be dispensed with, as can lights (robots work well in the dark). Best of all, robots never get tired, aren't mistake prone, and don't go on strike or ask for pay raises.

Robots are the new industrial revolution. Japan is the new Victorian Britain. Because Japan can make higher quality goods at less cost than any other country, they are prospering just as Victorian England did under similar circumstances a century ago. Like the British before them, Japanese tourists and businessmen seem to be everywhere. And just as machines previously spread to those countries which were prepared for them, so to are the robots spreading.

The vast stores of information accumulated by civilizations and science suggest a starkly different outcome from boom and bust. Instead of restraining ourselves to fit in with the other life forms, we will continue to restructure other life and the planet's ecosystems to suit our own civilizations.

Humanity was the breakthrough life needed to reach a new, higher level of complexity, to move beyond insect superorganisms to effective empires and a true planetary superorganism. As with previous breakthroughs, this one involves merger, in this case between humans and machines.

Until recently, the idea of civilizations dominating the planet and transforming it for the benefit of humanity was considered appropriate, a positive goal for us to work toward. As this goal has been transformed into reality, however, a few nervous nellies are beginning to have second thoughts. If we have indeed entered a new era of life on this planet, if a new evolutionary force has become predominant, then whether or not a few humans question the desirability of it all won't make any difference. The forces driving the evolution of life are greater than those of human opinion or desire. No ant would be so foolish as to think she had any real influence on her colony. Perhaps it is time we humans stop pretending that we have any influence on human civilizations (or for that matter that we ever had any real influence). We are cogs, little pieces in big superorganisms -- and always have been since the first civilization. But that's okay. Civilization has become a real winner.

The first DNA life may have totally taken over life on Earth as it then existed -- certainly no trace of non-DNA life remains. But highly successful DNA life was soon superseded by other, fancier DNA life, and so it has gone ever since. It may be very much the same with human life and civilizations. One senses that cultural evolution has always been impatient with our chimpanzee selves. The first human superorganisms only came into existence when we figured out a way to circumvent our chimpanzee genes and trick the masses into working all day long most every day. Nor did civilizations really take off until inefficient human labor was replaced with effective machine labor. It is fossil-fuel machines, not humans, that are responsible for civilizations really taking off. With computers and global networking, we sense the continued impatience of cultural evolution with human limitations; this time with mental as opposed to physical limitations.

Slowly but surely, humans are becoming less relevant as cultural evolution gathers strength. We are gradually being eased out of the way. Our replacements still need us -- they cannot yet reproduce without our help. So the symbiosis of human and machine will continue -- at least for a while. We are, however, rapidly becoming the junior partners in this arrangement. Someday we may not be needed at all. Although genetically we might end up without descendants, artificial machine life will, in the title of Hans Morevec's book, be our *Mind Children*.

If any genetic life remains, it will be because artificial life wants it to for one reason or another (extensively reengineered of course) or because some weeds and viruses just can't be exterminated. One suspects the latter will be the case, that artificial life will have more than just software viruses to contend with.

This future for life on planet Earth is completely unrestrained. Life will continue its relentless quest to get around all blocks to the accumulation of information and more efficient use of energy and material resources to create complexity. In looking back on the history of life on Earth from some future vantage point, our mind children will note that in our current time, DNA life had, with insect superorganisms, gone about as far as it could. The growth of complexity had come to another standstill until the era of human cultural evolution began. It quickly led to real culture. Machine culture.

Instead of being replaced we might just be totally reengineered. Taking charge of genetic evolution, cultural evolution, with generous help from computers, could reengineer many life forms, including ourselves into more orderly, logical, and efficient versions. Thank you Mr. Spock! Simultaneously, cultural evolution could shrink computers to biologically small sizes, fostering nano-tech self-reproduction, Marvin Minsky style, in itsy bitsy factories. Eventually, as suggested by Hans Moravec, the line between culturally reengineered genetic life and cultural designed-from-scratch robotic "life" would blur, would be forgotten as our mind children, although not our genetic children, propagate into the distant future.

We humans should not be sad about our coming replacement or alteration beyond recognition. It is a great honor to be the sole, single-species link between two major ears of evolution, genetic and the cultural. Only the original DNA life has been similarly honored, linking the chemical and genetic evolutionary eras. As Marvin Minsky suggested, "We humans are just dressed up chimpanzees. Our task is not to preserve present conditions, but to evolve, to create beings better, more intelligent than we."

Even if we are superseded in the stream of evolution, we may be physically or at least informationally preserved. Just as we now set parks aside to preserve life that can no longer successfully compete, so our machine or designer descendants may set a few of us old fashioned, inefficient humans aside in a new Garden of Eden, protected (by flaming Cherabims) from a world we helped create but are no longer fit enough to survive in on our own. Eden, a small portion of a planet set between two rivers will preserve the last of these strange life forms that naturally evolved instead of being purposely designed.

In the long corridors of time, species fade away, entire planets are left behind. Even if we *Homo sapiens* were protected for a while in Eden, we too would pass. Our DNA sequence, however, may be preserved in the great bank of accumulated information that will remain the heart of the cultural era. We know, already, whose DNA information will be passed on to eternity, the individual chosen to represent humanity forever. It is the African-American, Henrietta Lacks. In 1951, when she was thirty, she was diagnosed with an unusually aggressive cancer that killed her later that same year. A researcher at Johns Hopkins University found that cells from her biopsy sample grew well in vitro, not stopping their growth after the 50 or so replications that most cells do. Named after the first two initials of the donor's first and last names, these "HeLa" cells kept on duplicating and duplicating, hundreds and then thousands of generations. A mutation had created the ideal laboratory human cell. Samples rapidly spread to laboratories throughout the world. Soon HeLa cells became the de facto human DNA standard. As robotic DNA laboratory machines began their analysis of the human genome, the computer library of DNA sequences they began building up were those of Henrietta Lacks. HeLa cells have even been sent into space.

In the long stretch of cosmic time, it seems likely that another comet will eventually strike the Earth, setting the entire planet back to a simpler, less diverse state. Like the dinosaurs, we, along with all other large, complex life, will become history. With the dawn of the cultural era, however, this is no longer the necessary fate of Earth's life. When the next comet heads directly for Earth we will be ready. Instead of the planet being wiped out with the force of 10 thousand human nuclear arsenals set off all at once, a rocket will intercept the comet and, by exploding a super H-bomb beside it, nudge it off its collision course. Other life, if it could only understand and speak, would thank humanity -- especially Edward Teller -- for preserving the planet's remaining diversity.

Confined to Earth, all life would eventually perish as our sun turned into a red giant, consuming or at least melting our planet. But long before this happens, our mind children will have spread through the galaxy. Already we have taken the first faltering steps beyond our cradle. Ignoring the prophets of doom, we reach for the stars.

FURTHER READING

Botkin, Daniel B., 1990, *Discordant Harmonies: A New Ecology for the Twenty-First Century.*

Colinvaux, Paul, 1978, *Why Big Fierce Animals Are Rare: An Ecologist's Perspective.*

Kennedy, Paul, 1993, *Preparing for the Twenty-First Century.*

Lewis, Martin W., 1992, *Green Delusions: An Environmentalist Critique of Radical Environmentalism.*

Moravec, Hans, 1988, *Mind Children: The Future of Robot and Human Intelligence.*

Simon, Julian L., 1981, *The Ultimate Resource.*

Worster, Donald, 1994, *Nature's Economy: A History of Ecological Ideas* (2nd ed.).

CHAPTER 11

SUSTAINABLE BIOSTARS
Modest Self Restraint avoids the Crash

Reach for the stars? How foolish! We are confined to Earth and its vicinity and will be for the foreseeable future. This is our home, and we better stop treating it like a crash test dummy. Love it or loose it! Unbridled competition and consumption are a sure recipe for disaster.

We have won our struggle with nature. Now we must face the dangers of victory. If we were a normal out-of-control species, our future would be a bust even more spectacular than our boom. If we were a normal human civilization we would expect to collapse to a simpler, less populous state. We know, however, about boom and bust life. We know about the collapses of previous civilizations -- why they collapsed. We know, in short, not only how we came to be, but how we came to be in our present straits. This information could give us an unprecedented chance to actively create one of any number of alternative futures. In this and the final chapter we consider two of these alternatives.

Respectable numbers of humans in every civilization now know how we came to be. They have a decent grasp of the consequences of our current actions. They realize that unrestrained growth can only lead to a bust. This awareness could work magic. We might even restrain ourselves!

Self-restraint by any life form would certainly be unexpected, entirely without precedent. It would, without doubt, be biologically perverse. Ants, through no virtuous self-restraint on their part, were checked by other life before they got suicidally out of hand. They, thanks to others, have achieved the rare distinction of becoming sustainable biostars; a distinction ecologically possible for only a few forms of life on any given planet. Like the ants, we too could become sustainable biostars if we ourselves furnished the restraint. In so doing we might achieve a balance between ourselves and the rest of the planet while still diverting the lion's share of resources to our own ends.

In the long history of life on this planet, no species has ever controlled its own numbers or limited its own biological success voluntarily. On the other hand, no other species has ever accumulated so much information, understood where it came from or where it is likely to go or, for that mater, ever cared. Instead of going with the flow and crashing to oblivion, we might act on our understanding of our evolutionary predicament. We could--for the first time ever -- become self-restraining life. Although this would eliminate all biological precedent, it would open up a multitude of possible futures.

It seems an ironic coincidence, after five thousand years of civilization, ten thousand years of agriculture, and four billion years of life (all without the slightest hint of true understanding or any self-restraint), that modern human civilization (having finally understood its predicament) has only a hundred years or so to take crucial, effective, and precedent-setting action to avoid catastrophe. It is not a coincidence. The same accumulated information that led us to understand our predicament also led us to the machines and the fossil-fuel bonanza that generated immense food surpluses and, via science, to the improved sanitation and medicine that created the surge in population that is now eating these surpluses and threatening us all.

Our window of opportunity may be a narrow one. Once a collapse occurs, it could be so severe that we might enter a dark age from which our kind would never emerge -- at least with our civilizations and scientific knowledge intact. There is a paradox in all this. On one hand, machines, easy fossil fuels, and newly-discovered, sparsely-populated continents fueled a golden age that sustained and encouraged the scientific discovery of ourselves and our true evolutionary predicament while staving off collapse. On the other hand, should we fail to take advantage of the knowledge of our predicament gained during this temporary reprieve, the collapse, when it finally comes, will be all the more severe.

It is certainly possible that our self knowledge may have come to late. We may have damaged the ecosystem beyond repair, setting in motion ponderously-moving planetary forces that we cannot begin to control. The inertia of what we've already done may carry us beyond the point of no return, even if we mended our ways immediately. Even if there is still time, we may be unable to effectively restrain ourselves. Human culture may have an inertia of its own. It may be resistant to effective control. In either case, the window of opportunity, instead of being an opportunity might be the crowning touch to the human tragedy. Knowing what was coming, we could be powerless to stop it. This slow-motion nightmare would last a few hundred years until we crashed to oblivion. Let us assume, however, that we are not too late, that we can exercise effective self control.

Many degrees or amounts of self control might be possible. There is a wide range of potential futures in which we might exercise self control, but not enough to avoid a crash. A unique future is one where we apply just enough control to avoid collapse, a mini-control future. Many astute observers of the human scene feel we will be lucky to achieve even this. They warn against aiming for impossible utopias. Aiming too high, they suggest, would only increase the likelihood of a catastrophic collapse that would be the end of humanity. We should be practical. There are many limits to what can be done. If we are lucky and pull together, we might -- just barely -- avoid a crash.

To avoid an imminent collapse, we need to satisfy the health needs of the Earth's biosphere. Human activities must be made ecologically sustainable. This requires a new international economic order that gives first priority to an evolutionary novelty, the first true environmental ethic. It also requires that we stabilized human population.

As desirable as a healthy ecosystem would be, it may not be possible. Proposals for economic under-development, for slowing down and living in balance with a finite world, may not be practical due to the close link between economic and military power. Unilateral economic deceleration by any major power would be tantamount to unilateral disarmament. This means that countries cannot, individually, return to lower levels of economic activity. This is humanity's Catch 22: If any country opts out of the economically fatal spiral toward collapse, it will be taken over by a neighbor or a dominant world power and shoved back into the race. Yet if a country continues onward, it will collapse with all the others. So the fatal competition continues. Human population and its consumption spirals ever upwards regardless of the costs, ecological or human.

We need a planetary superorganism.

The ecological problems facing us are planetary in scope and require a coordinated planet-wide effort for their resolution. There is hope, for all around us we can the beginnings of a global system of governance, even a global culture (Hollywoodish as it may be). Increasingly, the entire planet is operating under one economic system. The tentacles of television and Internet are growing stronger day by day.

On the other hand, international corporations, individual governments, and military establishments have vested interests that promote economic growth and endless military competition. These organizations appear to be in control of humanity. They make the key decisions, determine policies, and decide our future and that of the planet. As Stephen Boyden wryly noted, "We have recently left behind the absurd notion that humankind's task is to conquer nature. It is culture, not nature that has to be subdued."

In the past, the masses have been easily duped by clever propaganda and half truths. However, with the advent of widespread, almost instantaneous communications, increasing universal education, books, and libraries, those in control are being forced to increasingly pander to the interests of the controlled. As information has leaked, even slipped wholesale from rulers to subjects, an opening has been created for humanity to nudge cultural evolution in a new direction.

With growing scientific understanding of the likely future consequences of our actions, national governments are finding it increasingly difficult to justify short-sighted decisions that jeopardize our future. The legitimacy of governments that fail to properly assess the long-term carrying capacities of their national territories, to keep populations and resource utilization within sustainable bounds, are being questioned. Politicians are beginning to get the message.

Rather than abandoning capitalism, governments are creatively using new taxes and financial incentives to make environmental consequences felt within the short-term horizon most real-world companies operate within. This is shoring up the near-sighted weakness of capitalism (and of evolution for that matter). The combination of an informed public and far-sighted governments could create an environment where doing the right thing would also be profitable.

Over thirty countries now have stable populations. These countries are, in the main, industrialized countries where the education of one's offspring is expensive yet vital for success, where the well-being of the old is assured through social security, not through having many offspring. In these countries children are a financial liability, an economic luxury affordable only in small quantities.

Most of the world's population is still growing rapidly, however. Plentiful children remain, in many countries, the only old age security. Just as industry can be incentivised to do the right thing for the long-term benefit of the planet, so we can incentivize people to have fewer children by providing old age security pensions, universal education, and assuring women equality and economic enfranchisement. It would be in the long-term best interests of industrialized countries to finance these advancements in poorer countries.

Like it or not, we are all, rich and poor, in this together, passengers on the same spaceship. Like it or not, we humans are now in this spaceship's driver's seat. The future of life on this planet will, at least for the next several million years, mainly depend on what one species does, on how it decides to utilize its vast accumulation of information, on whether or not it decides, contrary to all its instincts and the history of life, to voluntarily restrain itself.

FURTHER READING

Brown, Lester R., 199?, *Who Will Feed China?*
_____, 1996, *State of the World.*
Commoner, Barry, 1975, *Making Peace With the Planet.*
Heilbroner, Robert, 1995, *Visions of the Future.*
Rosenzweig, Michael L., 1974, *And Replenish the Earth: The Evolution, Consequences, and Prevention of Overpopulation.*

CHAPTER 12

CHIMPANZEE PARADISE
High-tech Garden of Eden

A planet dominated by 5, 10, or even 20 billion humans may be possible, even sustainable, but who in their right mind would ever wish for such a monstrosity? Our lovely forests and meadows would be gone forever, replaced by a boring sea of humans and their domesticated plants and animals. The once rich diversity of life, tens of millions of different species, would be replaced with a dreary sameness. Our blue and green globe that once sailed with beauty through the inky blackness would become a hog pen with standing room only for the victorious species, *Homo sapiens*. Pigship Earth.

Humans and their necessary service species, would not be the only remaining life, however. With the entire planet one big disrupted, continuous human farm and city, insects, rats, and microbes that have always benefitted from civilized disruption would have a field day. The planet would become a battleground between humans and their domesticates on one side, and weeds, vermin, and disease on the other. This dreary biological warfare could drag on until some lucky microbe finally broke the code to the largest-ever single-species biomass jackpot: humanity. Packed together like sardines, people jetting daily about the entire planet, we would be gone in a flash. Lacking such a lucky microbe, we could only pray, as our boring existence stretched on and on, that some gigantic meteor would smash Earth again, putting us out of our misery.

If we are capable of self restraint, why would we apply a minimal dose of it, just enough to avoid catastrophic collapse but not enough to avoid a living Hell? If we are going to be the first life to restrain itself, why not do it right, go all out and *really* restrain ourselves? Self-restraint opens up a virtual infinity of unnatural futures. In the last chapter we considered one extreme, the minimal restraint required to avoid an impending bust. In this, the book's final chapter, we consider the other extreme, self-control so complete that we could create a paradise on Earth. Given our druthers, what should we aim for?

As Randolph Nesse and George Williams pointed out in *Why We Get Sick: The New Science of Darwinian Medicine*, our bodies and minds evolved, over millions of years, to serve lives spent in small groups hunting and gathering on the plains of Africa. Natural selection hasn't had time to revise our bodies or rewire our brains for civilized life under conditions radically different from those which prevailed in prehistoric East Africa. If we are to meet individual human physical and mental health needs, greater recognition needs to be given to the many ill effects that civilized humans are currently subjected to. We need to restructure civilization to treat its members more as chimpanzees and less as ants. In a rational world, we should reap the benefits of civilization without its frightful consequences. We must create an environment in which humans will be physically, mentally, even socially healthy.

Hunting and gathering were varied activities enjoyed by our hunter-gatherer ancestors, providing them with beneficial exercise while allowing them to make direct and gratifying contributions to the welfare of their group. These were not overly time-consuming activities; they allowed generous time for naps and social gatherings. The wild game our ancestors ate was naturally low in fat, the vegetables they consumed high in fiber. Hunter-gatherers rarely put on weight as they aged, nor did they suffer from high blood pressure. Cardiovascular difficulties were rare, as were cancers. Hunter-gatherers were generally relaxed. They didn't hunt or gather more than they required to eat. The modern concept of always being productively busy is foreign to hunter-gatherers. The suggestion that work is virtuous is a recent, self-serving bill-of-goods that civilization's elites sold to the masses (although they often failed to apply it to themselves).

Civilizations, while they benefitted elites, were not healthy for the masses in earlier times. Civilized diets lacked variety and usually lacked sufficient calories. The work was long, hard, and repetitious. As Mark Nathan pointed out in *Health and Civilization*, with the advent of civilizations and its accompanying malnutrition, humans became noticeably shorter, losing some six inches in average height. Humans were able to adapt to the harsh conditions of civilization, but as Stephen Boyden points out, "The frightful threat posed by human adaptability is that it implies so often a passive acceptance of conditions which really are not desirable for mankind."

Throughout the long history of civilization, the most common diet problem has been underconsumption, the most common physical problem overexercise. Our modern era, driven by fossil-fuel machines, has produced food surpluses and freed much of humanity from agricultural labor. The most pressing health problem in industrialized nations today, is one of over consumption and under exercise -- problems formerly restricted to the elite. Fat, sugar, and salt, rare and difficult to obtain commodities in our hunter-gatherer past, have suddenly became abundant.

We are, via science, learning what is conducive to human health and what is detrimental, both physical and psychological. We are learning that which humanity has become genetically adapted too over the ages is generally healthy. Conversely, that which is novel and new, that which we have not had time to genetically adapted, is often unhealthy. There are exceptions in both directions. Studying and understanding the healthy aspects of hunter-gatherer lifestyle does not imply we must return to it to reap its benefits. Hunter-gatherer ancestors did not have the benefit of antibiotics, tetanus shots, anesthetics, plaster casts, glasses, or false teeth.

It is clear, however, that humans need the hunter-gatherer virtues of a varied diet, frequent and varied physical exercise, variety in daily experience, and a sense of personal involvement and belonging to small groups of family and friends that are increasingly valued over the years. As a social animal, we need our mates, children, relatives, and friends; even our pets. We require intimate, long-lasting relationships. We need community. We do best when changes are slow-paced; when grandparents and grandchildren don't live in radically different worlds. We gather psychic strength from a close and harmonious coexistence with nature. It is not surprising that we look back nostalgically to the Garden of Eden, for we have never forgotten our true home.

A healthy, sustainable future does not imply a non-technical future. It would be disastrous for us to abandon science or technology at this crucial juncture. Science can show us how to heal the Earth's ecosystems. New technologies are needed, such as solar power, to make our presence more ecologically benign. More of the wrong sorts of science and technology could be disastrous, but so too could an insufficient dose of the right ones. Science and technology ushered in our current age of plenty. We must turn to them not only to bring their benefits to all humanity (greatly reduced in numbers in the process), but also to eliminate the harmful side effects of science and technology themselves. Fight fire with fire, knowledge with knowledge. We must depend on science, our most reliable source of knowledge, to convince humanity of its peril and the nature of its salvation.

Until recently, it would have been difficult for culturally-varied humanity to agree on much of anything. Science now provides us with a unified, cross-cultural view of humanity and the world, of who we are, how we came to be, and our alternatives for the future. We should not blame ourselves for not yet being fully united in this respect. The last few crucial pieces of science's evolutionary view of humanity have only recently clicked into place. We didn't even understand the basic nature of life until James Watson and Francis Crick discovered the structure of DNA in 1953. It was in the 1960s that Jane Goodall began her observations of chimpanzees in the wild. Only in the 1980s and 1990s have the natural and social sciences begun their unification by way of a joint evolutionary view of Earth and humanity.

The pieces are now in place.

What is needed is for individuals everywhere to come to understand the human situation, to understand ourselves from a unified human, biological, and ecological evolutionary perspective. If the basic understanding of humanity and the planet provided by our scientific, evolutionary views became widespread, then a change in the dominant outlook could occur. This could lead to changes in corporate, governmental, and military policies compatible with the health of humanity and the planet. Humanity can no longer consider itself apart from the rest of life on this planet.

An increasing number of people are expressing environmental concerns.

While this demonstrates that their hearts are in the right place, compared with the scale of the problem most measures are still cosmetic. While the conservation movement has raised public awareness and won a few small victories, we are being overwhelmed by a tidal wave of destruction. As more of us learn of our true biological situation, however, perhaps together we can effectively channel the future course of cultural evolution in a more desirable direction. The most important action of all may be to educate the public as to the nature of our present circumstances and future prospects.

Our closest relatives, the chimpanzees and bonobos (as well as the other great apes, the gorillas and orangutans) are all headed towards extinction as the last of the great equatorial jungles are chain-sawed and bulldozed to oblivion. This should give us pause. Will we be next? The present diversity of life took 65 million years to build up since the Earth was last struck by a mega-meteor. If the present human-perpetuated mass extinction is not halted, diversity of life will plummet again. It would again take millions of years to recover.

Having promoted ourselves to Masters of the Planet, we owe it to other life, to the planet, and especially to ourselves to take charge of that rogue ape -- humanity -- and preserve the biological diversity that remains. We need to switch from being the planet's chief plunderer to being its guardian.

We must become Gaia.

One might think that it would be best if humanity rejoined nature as quickly as possible. With our present billions, that would devastate what little non-human nature remains. Quite the contrary, we must separate humanity from what is left of nature. We must quarantine this dangerous species from other life. We are now starkly different from all other life -- truly unique -- and will remain so. The human cultural genie cannot be stuffed back into nature's genetic bottle. Our presence will forevermore be unnatural and have to be controlled. Having eaten the fruit from the Tree of Knowledge, we cannot return to the Garden of Eden without careful supervision. As our numbers plummet, we may be able to increasingly enjoy a rejuvenated Eden, but we will never be able to return it as innocent Adam and Eves. Even if we could leave our science and technology behind, without careful self-supervision the take-over would only happen again. We must become and remain a carefully organized and controlled planetary superorganism that watches over the Earth with intelligent foresight and constant vigilance.

From our current vantage point, it is difficult to envisage a future where there are only millions of humans instead of billions, a future where progress itself has slowed to a crawl. Throughout most human history, the change from one generation to the next was always imperceptible. Progress was a foreign concept. If anything, the past was perceived as better than the present; we had departed downward from a previous golden age.

The idea of progress, fueled by science and technology and nurtured by capitalism, as being good, even inevitable, is a recent concept launched by the likes of Francis Bacon and Adam Smith. It reached its Victorian peak prior to World War I. With atomic bombs, ozone holes, and a planet clearly overcrowded, we now view "progress" as, at best, a two-edged sword. But could "progress" have ever been more than a transient phase? Science, after all, can only make serious progress as long as there are major discoveries left to be made. While important discoveries remain, there is a falling off in such discoveries already. Capitalism, which prospers from science, technical progress, and expansion, must also eventually subside. Infinite expansion is simply impossible. From a vantage point far in the future, the era of major scientific discoveries, rampant technology, and unbridled capitalism will be seen as a transient phase of sudden, unusual change.

The great gift of this transient phase will be a universal age of plenty. Before, only the elites of civilization led the good life, but after -- thanks to the hard-working, productive, robotic machines and bountiful new sources of clean energy -- all humanity (the optimum few hundred million of us that would remain in the new era), will live the good life of the elite.

We might view the entire past 10 thousand years of agriculture, civilization, and science as a short, hard-working, turbulent phase between two well-adapted, leisurely eras. We humans naturally like to do what we are fit to do, what we evolved for millions of years to do. These do not include hard work in the hot sun (or hard, monotonous work of any sort in factories or offices). We are chimpanzees, not ants.

Freed from the necessity for work, from the anxiety of constant change, from economic insecurity, we will do what humanity has always done under such circumstances; relax and enjoy ourselves. As Gunther Stent suggested in *The Coming of the Golden Age*, it has begun already. The hippies and beatniks are leading the way for the rest of humanity. Music, sex, and peaceful vibes will be in. Work, competition, and war will be out.

Without the fear of being constantly blindsided by new scientific revelations, myths and religions of all sorts will prosper as never before in an explosive profusion of human creativity. Our imaginations always exceeded mere objective reality. Science's increasingly dated major revelations will -- all of them -- soon take on the sort of excitement we currently show for the revelation that the Earth is round. As all religions take science as an obvious, boring given, that which has always intrigued humanity -- our imaginative stories, speculative new ideas, creative art and music -- will flower in a profusion of local and personal diversity.

Stent viewed the coming golden age as similar to Polynesia, a contended humanity in a land of easy plenty. Far from being unchallenged and without a mission in life, however, the citizens of the coming golden age will have their hands full as the overseers and caretakers of an entire planet. They will be occupied for many centuries restoring the planet from the ravages of the transitional phase, not to mention bringing human population down to its optimal level.

As guardians of the planet we will, presumably, defend the Earth against planet-busting meteors. Most of all, however, we will defend the planet against ourselves. We will insist on becoming and remaining responsible citizens of planet Earth. In this task we will be greatly aided by our newly unified scientific understanding of life and humanity, of how we came to dominate the planet, and how, at the last minute, we avoided disaster by taking firm charge of ourselves and our only home. In this we owe a great debt to the environmentalists who have led the way. Lest we slip back into our old, irresponsible ways again, the young will forevermore be inoculated by learning the cautionary tale of *The Chimpanzees Who Would Be Ants*.

FURTHER READING

Nesse, Randolph M., and George C. Williams, 1994, *Why We Get Sick: The New Science of Darwinian Medicine.*

Stent, Gunther S., 1969, *The Coming of the Golden Age.*

FUTURES MOST LIKELY AND DESIRABLE

I occasionally present my unified scientific story of humanity as a one-hour talk at high schools and universities. Afterwards I ask the students which of the four futures do they consider most likely? They invariably elect *Boom and Bust* by an overwhelming, near unanimous majority. That anyone would consider any other future as likely strikes them as unrealistic wishful thinking. With respect to recent environmental initiatives, they seem to think they are too little, too late. Some suggest they are mainly a multinational PR blitz while they finish raping the planet.

When, for the first time, I asked which future is most desirable, I expected they would all vote for *Chimpanzee Paradise*. They didn't. The vote was split between *Robotic Triumph* and *Chimpanzee Paradise*. Why? Some want to continue our exciting, Star Treckish adventure as Masters of the Universe, while others would rather see us rejoin nature as ordinary citizens. Several that voted for *Robotic Triumph* suggested that *Chimpanzee Paradise* would be boring, somewhat mindless. How, in any event, they all asked, would we ever get our population down to millions without a bust or unpalatable raw force?

Turn about, they insisted on my views. Here, for what it is worth, I repeat my views: I contend that a crash is unlikely--at least one that eliminates *Homo sapiens* altogether. Africa seems to be crashing, but without taking the rest of the planet with it. We are a tenacious species and ecosystems are tough. Nor is *Pigship Earth* probable -- at least not for very long. With twenty billion humans, some lucky microbe would soon break the code to the ultimate biological jackpot, creating a spectacular crash.

Although *Chimpanzee Paradise* may indeed be just a utopian fantasy, *Robotic Triumph* would continue life's unbroken record of information accumulation, advancing itself to a new high in the hierarchy of complexity. Considering the past track record of life and the current wild burst of energy shown by cultural evolution, it appears to me that a breakthrough has been made that ranks right up there with the first DNA life or the first photosynthetic life. We are living in the midst of a massive radiation, a virtual explosion of human castes (specialities) and machines. I don't think we will be taken over by machines, but I don't think we will be able to do without them either. After all, we have been partners with tools since we first stepped foot on the African savanna over 2 million years ago, and our current prosperity is due -- almost entirely -- to fossil fuel machines, the direct descendants of these first stone tools.

My most desirable future? While *Robotic Triumph* is tempting (I've always loved robots), warp drive is pure science fiction (sorry about that, Captain Kirk). We will, at least for the near future, be effectively confined to Earth. Furthermore, we will always be chimpanzees at heart, physically and psychically unfit for any other life. So I cast my vote for *Chimpanzee Paradise*, the one future that restores and safeguards our only home. It's a long shot, but let's chimpishly rebel against being mere ant-like stepping stones to robots. Perhaps, with the planet secure and the rogue ape under control, we could -- taking our time -- build a starship or two on the side.

Russell Merle Genet

Genet dissipated his teenage years searching through science and history for human meaning in a vast universe. Summer vacations were spent camping, hiking, and reading. A morbid fascination with self-acting machines launched a continual stream of model airplanes and rockets from his basement lab. Undergraduate education was channeled into both the human (psychology) and the robotic (electrical engineering), while his graduate degree combined the two (the analysis of complex man-machine systems). Genet worked for some time as, quite literally, a rocket scientist, developing automated guidance systems. Then he drew on his expertise in airplanes (he is a pilot), computers, and psychology, to develop flight simulators (giant video games) for training pilots. His cosmic, scientific hobby for many years was astronomical research -- measuring the brightness variations of eclipsing, pulsating, and spotted stars. Inherently lazy, he automated his telescope, sleeping while his robot observed. Genet has authored or edited a dozen books on computers, robotics, and astronomy. Retiring from research and development at fifty, he eagerly but foolishly renewed his teenage quest for human (and robotic) meaning in a vast universe. Now on a never-ending "teenage" summer vacation, gypsy Genet spends "winter" camping on the beach in the south Pacific, and summer in the Arizona mountains. For the amusement of the idly curious, he summarized the latest scientific views of humanity (and robots) into this, his most recent book.

INDEX

DATE DUE

GAYLORD   PRINTED IN U.S.A